高等院校计算机技术"十二五"规划教材

Linux 程序设计实践与编程技巧

主 编 刘加海 骆建华
副主编 王晓池 杨 锆

浙江大学出版社

内容简介

本书是浙江大学出版社出版的《Linux 程序设计》一书的辅导书,书中内容主要分两大部分。一是程序设计要点与技巧;二是实践部分,包含 16 个实验,每个实验有 6~7 个任务。本书包含:终端基本命令、Shell 程序设计、Linux 环境下 C 程序编译技巧、Linux 环境下 C 语言系统函数的应用、Linux 环境下文件的 I/O 操作、进程的控制与进程调度、线程、Linux 环境下的图形与游戏程序设计、网络程序设计、字符驱动程序设计、串行程序设计的编程知识要点,编程技巧与程序设计实例。并包含以上各部分及 Linux 环境与 Windows 环境资源共享设置的实验。

本书重点给出 Linux 程序设计的要点与技巧,希望能够为本科生、研究生、嵌入式工程技术人员、Linux 程序设计师及 Linux 程序爱好者提供有效的帮助。

图书在版编目(CIP)数据

Linux 程序设计实践与编程技巧,刘加海,骆建华主编.
—杭州:浙江大学出版社,2013.6(2018.3 重印)

ISBN 978-7-308-11315-1

Ⅰ.①L… Ⅱ.①刘… ②骆… Ⅲ.①Linux 操作系统–程序设计 Ⅳ.①TP316.89

中国版本图书馆 CIP 数据核字(2013)第 060654 号

Linux 程序设计实践与编程技巧

主　编　刘加海　骆建华
副主编　王晓池　杨　锆

责任编辑	武晓华
封面设计	刘依群
出版发行	浙江大学出版社 (地址:杭州市天目山路 148 号　邮编:310007) (网址:http://www.zjupress.com)
排　　版	杭州理想广告有限公司
印　　刷	浙江省良渚印刷厂
开　　本	787mm×1092mm　1/16
印　　张	20.25
字　　数	520 千
版 印 次	2013 年 6 月第 1 版　2018 年 3 月第 2 次印刷
书　　号	ISBN 978-7-308-11315-1
定　　价	39.00 元

版权所有　翻印必究　印装差错　负责调换

浙江大学出版社发行中心联系方式:0571-88925591;http://zjdxcbs.tmall.com

前 言
PROFACE

随着 Linux 操作系统在各个行业的广泛应用，企业对 Linux 人才的需求也将持续升温，具有 Linux 专业知识及开发经验的人才越来越受到用人单位的欢迎。

本书是浙江大学出版社出版的《Linux 程序设计》一书的辅导书，书中内容主要分两大部分：一是程序设计要点与技巧，二是实践部分。第一部分是编程技巧，包含了 Linux 基本命令操作、Shell 程序设计、makefile 工程文件、Linux 环境下 C 语言系统函数的应用、Linux 环境下文件的 I/O 操作、进程的控制与进程通信、线程、Linux 环境下的图形程序与游戏程序设计、网络程序设计、串行程序设计等内容的要点、应用技巧及程序设计实例。实践部分包含 16 个实验，每个实验有 6~7 个任务，含终端基本命令、Shell 程序设计、Linux 环境下 C 程序编译技巧、Linux 环境下 C 语言系统函数的应用、Linux 环境下文件的 I/O 操作、进程的控制与进程调度、线程、Linux 环境下的图形与游戏程序设计、网络程序设计、字符驱动程序设计、串行程序设计、Linux 环境与 Windows 环境资源共享设置。

十多年来，笔者一直在浙江大学计算机学院、浙江大学软件学院、LUPA 全国 Linux 师资培训中心讲授《Linux 程序设计》这门课程，此课程深受本科生、研究生、高校教师的欢迎，此后笔者根据多年的教学经验编写了《Linux 软件工程师（C 语言）实用教材》一书，此书由倪光南院士写序，在读者中深受欢迎。应广大读者的需求，笔者根据教学过程中发现的问题及教学的需要，编写了《Linux 程序设计》一书，此书由浙江省教育厅推荐为国家"十二五"规划教材。在使用中学生希望有本对于在实践方面与编程中遇到的难点给予指导的辅导书，所以又花了两年多时间编写了这本《Linux 程序设计实践与编程技巧》一书，书中部分内容选取浙江大学计算机学院、软件学院 2006—2012 级学生的作业，在此对这些同学表示感谢。希望本书能够对学习 Linux 程序设计的本科生、研究生、嵌入式工程技术人员及 Linux 程序爱好者提供帮助。

通过本书的学习，可以掌握以下内容：
- Linux 基本操作技巧；
- Shell 程序设计技巧；
- makefile 工程文件；
- Linux 环境下 C 语言编辑、编译、调试方示与技巧；

- Linux 环境下文件的 I/O 操作技巧；
- 进程的控制与进程调度技巧；
- 线程编程技巧；
- Linux 环境下的图形程序、游戏程序设计技巧；
- 网络程序设计技巧；
- 串行案例设计技巧；
- 典型案例设计技巧。

本书在讲解讨论内容时，首先给出实例，通过例子介绍程序设计的方法，通过大量的例子与清晰的程序流程使读者迅速掌握相关知识、编程技能与技巧，并通过大量的思考题帮助读者迅速提高程序设计能力。

本书中 16 个实验的设计，更是符合人们学习程序设计的心理特点，在每一个实验中首先给出程序的调试，在程序调试的基础上再根据给出的问题，对程序的关键语句学会自己设计，最后给出程序设计项目，培养读者完全独立设计程序的能力。如果比喻着学习"走路"，这相当于在每个实验中完成了让您看着别人走路、扶着您走路、让您独立行走的全过程，相信这 16 个实验会极大提高读者的程序设计能力。

本书由浙江大学城市学院刘加海教授，上海交通大学航空航天学院骆建华教授，浙江大学计算机学院季江民副教授，浙江大学城市学院严冰副教授、杨锆讲师、王晓池讲师，浙江大学宁波理工学院唐云廷副教授编写，全书由刘加海统稿。由于时间仓促及作者水平有限，书中难免存在疏漏和不妥之处，敬请广大读者批评指正，批评与建议请发到邮件 Ljhqyyq@yahoo.com.cn，以便及时修订。

目 录
CONTENTS

程序设计要点与技巧

第 1 章　基本命令　// 3

1.1　知识要点　// 3
  1.1.1　常用终端命令　// 3
  1.1.2　常用元字符*、?、~、[ ]的应用　// 7
  1.1.3　重定向符 |、>>、>、<　// 8
  1.1.4　单引号与双引号的作用　// 9
  1.1.5　文本编辑命令 vi 或 vim 的应用　// 10
  1.1.6　Linux 常用系统变量　// 11

第 2 章　Shell 编程　// 13

2.1　知识要点　// 13
  2.1.1　Shell 程序运行方式　// 13
  2.1.2　Shell 中变量的使用　// 13
  2.1.3　Shell 编程中参数替换　// 14
  2.1.4　Shell 编程中大段文字显示技巧　// 14
  2.1.5　随机数产生　// 15
  2.1.6　日期输出格式设置　// 15

2.1.7 在 Shell 中的算术表达方法　//16
2.1.8 Shell 程序设计中分支语句　//18
2.1.9 Shell 程序设计中循环　//19
2.1.10 Shell 中函数的格式　//20
2.1.11 Shell 中数组的使用　//21
2.1.12 文件与文件夹的判断　//22
2.1.13 某路径下文件总容量的判断　//23
2.1.14 菜单程序的框架　//23
2.2 程序设计实例　//24

第 3 章　Linux 系统 C 语言开发工具　// 32

3.1 知识要点　//32
   3.1.1 gcc 的使用　//32
   3.1.2 makefile 工程文件　//33
3.2 程序设计实例　//35

第 4 章　Linux 环境下系统函数的使用　// 37

4.1 知识要点　//37
   4.1.1 随机函数的应用　//37
   4.1.2 系统时间函数的应用　//38
   4.1.3 系统函数　//39
   4.1.4 数据结构中的函数　//40
4.2 程序设计实例　//41

第 5 章　Linux 环境下文件 I/O 操作　// 48

5.1 知识要点　//48
   5.1.1 文件操作　//48
   5.1.2 文件控制特性的判断　//51
5.2 程序设计实例　//53

第 6 章　进程控制　// 61

- 6.1　知识要点　// 61
  - 6.1.1　进程调度命令 at　// 61
  - 6.1.2　进程创建函数 fork　// 62
  - 6.1.3　僵尸进程　// 63
  - 6.1.4　wait 与 waitpid 函数　// 64
  - 6.1.5　僵尸进程的避免　// 65
  - 6.1.6　守护进程　// 65
- 6.2　程序设计实例　// 66

第 7 章　进程间的通信　// 78

- 7.1　知识要点　// 78
  - 7.1.1　Linux 进程间的通信方式　// 78
  - 7.1.2　进程间通信的特点　// 78
  - 7.1.3　管道通信的函数　// 79
  - 7.1.4　信号传送和处理　// 80
  - 7.1.5　消息队列应用　// 81
  - 7.1.6　共享内存函数 mmap 应用　// 83
- 7.2　程序设计实例　// 84

第 8 章　线　程　// 99

- 8.1　知识要点　// 99
  - 8.1.1　线程与进程　// 99
  - 8.1.2　多线程和多进程的对比　// 100
  - 8.1.3　线程中的常用函数　// 100
  - 8.1.4　线程中互斥锁的实现　// 101
  - 8.1.5　线程中信号量的应用　// 102
- 8.2　程序设计实例　// 103

第 9 章　网络程序设计　// 113

- 9.1　知识要点　// 113
    - 9.1.1　Socket 接口　// 113
    - 9.1.2　Sockaddr 和 Sockaddr_in 结构类型　// 113
    - 9.1.3　TCP 编程　// 114
    - 9.1.4　UDP 编程　// 115
    - 9.1.5　常用操作函数　// 116
- 9.2　程序设计实例　// 119

第 10 章　Linux 图形程序设计　// 134

- 10.1　知识要点　// 134
    - 10.1.1　SDL 库　// 134
    - 10.1.2　常用 SDL 库函数　// 135
- 10.2　程序设计实例　// 137

第 11 章　设备驱动程序设计基础　// 147

- 11.1　知识要点　// 147
    - 11.1.1　设备驱动程序概况　// 147
    - 11.1.2　字符设备驱动程序　// 148
- 11.2　程序设计实例　// 150

第 12 章　串行通信　// 162

- 12.1　知识要点　// 162
    - 12.1.1　串行通信　// 162
    - 12.1.2　串行通信程序设计流程　// 163
    - 12.1.3　串行通信程序设计步骤　// 163
- 12.2　程序设计实例　// 165

实践部分

Linux 程序设计实验报告 1——操作系统基本命令使用　// 175
Linux 程序设计实验报告 2——Shell 程序设计 1　// 180
Linux 程序设计实验报告 3——Shell 程序设计 2　// 185
Linux 程序设计实验报告 4——Linux 系统 C 开发工具　// 190
Linux 程序设计实验报告 5——Linux 环境系统函数的应用　// 198
Linux 程序设计实验报告 6——Linux 文件 I/O 操作 1　// 206
Linux 程序设计实验报告 7——Linux 文件 I/O 操作 2　// 212
Linux 程序设计实验报告 8——进程控制　// 224
Linux 程序设计实验报告 9——进程通信 1　// 233
Linux 程序设计实验报告 10——进程通信 2　// 240
Linux 程序设计实验报告 11——Linux 线程程序设计　// 251
Linux 程序设计实验报告 12——Linux 网络程序设计　// 259
Linux 程序设计实验报告 13——Linux 图形程序设计　// 267
Linux 程序设计实验报告 14——虚拟字符驱动程序设计　// 280
Linux 程序设计实验报告 15——Linux 串行通信程序设计　// 289
Linux 程序设计实验报告 16——Windows 与 Linux 操作系统间资源共享　// 303

程序设计要点与技巧

Linux程序设计实践与编程技巧

第1章 基本命令

1.1 知识要点

1.1.1 常用终端命令

1. 在 Linux 中，字符是区分大小写的。
2. 建立目录命令 mkdir

例如，在目录/mnt 下建立文件夹 usb

```
mkdir /mnt/usb
```

3. 在目录中搜索文件 find

```
find [path][options][expression]
```

常用参数：

—name：支持统配符*和?。
—atime n：搜索在过去 n 天读取过的文件。
—ctime n：搜索在过去 n 天修改过的文件。
—group groupname：搜索所有组为 groupname 的文件。
—user 用户名：搜索所有文件属主为用户名（ID 或名称）的文件。
—size n：搜索文件大小是 n 个 block 的文件。
—print：输出搜索结果，并且打印。

应用技巧：
(1) 根据文件名查找

例如，想要查找一个文件名是 lilo.conf 的文件，可以使用如下命令：

```
find / -name lilo.conf
```

find 命令后的参数"/"表示搜索整个硬盘，例如搜索/etc 路径下的文件名 smb.conf 应用参数。

```
find /etc -name smb.conf
```

(2) 根据部分文件名查找方法

例如，想要查找文件名中包含字符 abvd 的所有文件。

```
find / -name '*abvd*'
```

在/目录中查找所有的包含有 abvd 这 4 个字符的文件。

(3) 使用混合查找方式查找文件

例如，在/etc 目录中查找大于 500000 字节，并且在 24 小时内修改的某个文件，则可以使用-and(与)把两个查找参数链接起来组合成一个混合的查找方式。

```
find /etc -size +500000c -and -mtime +1
```

(4) 复制文件夹时，使用命令 cp，要加上参数-r，才能递归复制，不然只会将目录下的文件复制过来，而目录下的文件夹不会复制过来。

4.修改密码

Linux 终端里使用 passwd 命令修改密码可以很方便地修改密码：

```
[root@localhost root]# passwd
Changing password for your_username
New password: 新密码
Retype new password: 再输入一次新密码
Passwd: all authentication tokens updated successfully
```

如果忘记了密码，可以试图使用以下 2 种方法解决问题：

(1) 在系统进入单用户状态，直接用 passwd root 去更改；

(2) 用安装光盘引导系统，进行 linux rescue 状态，将原来/分区挂载做法如下：

```
cd /mnt
mkdir hd
mount -t auto /dev/hdaX(原来/分区所在的分区号) hd
cd hd
chroot ./
passwd root
```

5.文件权限设置命令 chmod

使用方式：

```
chmod [-cfvR] [--help] [--version] mode file...
```

说明：

Linux/Unix 的档案存取权限分为三级:档案拥有者,群组,其他。利用 chmod 可以控制档案如何被他人所存取。

mode：权限设定字串,格式如下：[ugoa...][[+-=][rwxX]...][,...]，其中 u 表示该档案的拥有者，g 表示与该档案的拥有者属于同一个群体(group)者，o 表示其他以外的人，a 表示这三者皆是。

+ 表示增加权限，- 表示取消权限，= 表示唯一设定权限。

r 表示可读取，w 表示可写入，x 表示可执行，X 表示只有当该档案是个子目录或者该档案已经被设定过为可执行。

-c：若该档案权限确实已经更改，才显示其更改动作。

-f：若该档案权限无法被更改也不要显示错误讯息。

-v：显示权限变更的详细资料。

-R：对目前目录下的所有档案与子目录进行相同的权限变更（即以递回的方式逐个变更）。

-help：显示辅助说明。

-version：显示版本。

例如：

将档案 file1.txt 设为所有人皆可读取：

chmod ugo+r file1.txt

或

chmod a+r file1.txt

将档案 file1.txt 与 file2.txt 设为该档案拥有者，与其所属同一个群体者可写入，但其他以外的人则不可写入：

chmod ug+w,o-w file1.txt file2.txt

将 ex1.py 设定为只有该档案拥有者可以执行：

chmod u+x ex1.py

将目前目录下的所有档案与子目录皆设为任何人可读取：

chmod -R a+r *

此外 chmod 也可以用数字来表示权限如 chmod 777 file

语法为：chmod abc file

其中 a、b、c 各为一个数字，分别表示 User，Group，及 Other 的权限。

r =4，w =2，x =1

若为 rwx 属性则 4+2+1=7；

若为 rw- 属性则 4+2+0=6；

若为 r-x 属性则 4+0+1=5。

例如：

chmod a=rwx file

和

chmod 777 file

效果相同

chmod ug=rwx,o=x file

和

```
chmod 771 file
```
效果相同

若用 chmod 4755 filename 可使此文件具有 root 的权限。

6.磁盘挂载

在 Linux 中所有的设备都以文件的形式显示在/dev 当中，对于老的 PATA 的硬盘，会以 hda 开头，对于目前采用比较多的 SATA 硬盘会以 sda 开头，光驱是以 cdrom 开头的，cciss 是惠普的磁盘阵列开头。

如果需要使用一个设备，可以通过 mount 命令来挂载设备。

例如：
```
mount -t 文件格式 /dev/cciss/c0d0p2 /home
```
对于 Linux 中基本是 ext 格式，较新的系统会采用 ext 4 格式，老一些不支持 ext 4 的版本会采用 ext 3 格式。

7.回收站的位置

回收站的位置用 "~/.Trash/" 表示。

8.下载命令 wget

可以下载互联网上面的文件到本地，格式为：
```
wget URL地址
```
例如：
```
$wget http://www.rarlab.com/rar/rarlinux-3.8.0.tar.gz
```

9.时间显示与时间设置的命令 date
```
date [-u|--utc|--universal] [MMDDhhmm[[CC]YY][.ss]]
```
查询：
```
[root@localhost root]# date
日 12月  9 07:09:14 CST 2012
```
修改：
```
[root@localhost root]# date 12242322.21
一 12月 24 23:22:21 CST 2012
```

date命令的使用：

直接输入 date 命令，将显示当前的系统时间。

使用 "date 月日时分年.秒" 的格式改变当前系统时间

使用 "date+格式" 按格式输出系统时间

常用的格式有：

%a,%A：输出星期几（%a 为简写，%A 为全称）

%b,%B：输出月份名（%b 为简写，%B 为全称）

%d：一个月内的第几天

%H：小时

%m：月份

%M：分钟

%N：纳秒数（可以用来生成随机数）

%S：秒
%Y：年
例如：
[root@localhost root]# date +%Y-%m-%d
2012-12-10

在 Linux 中有硬件时钟与系统时钟等两种时钟。date 可以改变系统时钟，而 hwclock 可以改变硬件时钟。例如：
[root@localhost root]# date
一 12月 24 23:46:36 CST 2012
[root@localhost root]# hwclock
2012年12月09日 星期日 07时35分19秒 -0.771641 seconds
[root@localhost root]#

思考执行结果：
t_date= $(date +%m%d)
n_date=$(date +%w)

10.检查磁盘命令 fdisk

检查磁盘，列出结果中有关/dev/sd的磁盘信息
/sbin/fdisk -l |grep /dev/sd

11.终止进程命令 kill

例如：在后台执行编辑命令vi，然后应用命令kill终止vi的执行。
[root@localhost root]# vi &
[2] 22633
[1]+ Stopped   vim
[root@localhost root]# kill 9 -22633
[2]+ Stopped   vim
Vim: CVim: 拦截到信号(signal) TERM
Vim: 结束.
[root@localhost root]#

1.1.2 常用元字符*、？、~、[ ]的应用

通配符是由 shell 处理的，它只出现在命令的"参数"里。当 shell 在"参数"中遇到了通配符时，shell 会将其当作路径或文件名去在磁盘上搜寻可能的匹配。若符合要求的匹配存在，则进行代换（路径扩展），否则就将该通配符作为一个普通字符传递给"命令"，然后再由命令进行处理。

符号	功能
[ ]	匹配[ ]中一个字符
{ }	在当前 shell 中执行命令，或实现扩展
?	匹配单个字符
*	匹配 0 个或者多个字符
~	表示主目录

例如：

列出 6-1.c 6-2.c 6-3.c 6-5.c 6-6.c 6-7.c 6-8.c文件的信息，可用以下命令：
```
ls 6-[1-9].c  -l
```

例如：

查找/usr/sbin及/usr/bin/两个目录中所有的C语言程序，可用以下命令：
```
find -name /usr/{sb,b}in/*.c
```

例如：

执行命令echo s{ab,cd}y，输出为saby、scdy
```
[root@localhost root]# echo s{ab,cd}y
saby scdy
```

echo函数可以用转义字符\来完成一系列的操作，但是实际使用的时候也许并没有这个效果。使用man来查看ehco，可以发现其默认的模式是忽略转义字符，如果在后面加上参数-e，就可以启用转义字符了。

1.1.3 重定向符|、>>、>、<

1.可通过'>'和'<'实现重定向输入和输出

Command<filename实现将command命令的输入变更为来自filename。

Command>filename 实现将 command 命令的输出输送到 filename 中而非显示器。在 shell 命令行的解析中，重定向操作按照从左到右的顺序进行。可通过文件描述符来实现输出和输入的重定向。

重定向符>>、>的区别是：应用>>时，filename 文件如已存在，则把结果添加到 filename 文件中，filename 文件如不存在，则创建文件；应用>时，filename 文件如已存在，则把结果覆盖 filename 文件的内容，filename 文件如不存在，则创建文件。

文件描述符0：代表一个程序的标准输入；

文件描述符1：代表一个程序的标准输出；

文件描述符2：代表一个程序的标准错误输出。

例如：
```
[root@localhost root]# cat < create.c > stdout.c
```

表示cat 命令的输入来自于create.c 文件，同时cat 命令再将数据传输出到stdout.c 的文件中。

例如：

`[root@localhost root]#grep "Joson" students 2>output.error`

表示在执行grep 命令中若有错误，则把其错误输出到output.error文件中。

例如：

`[root@localhost root]# sort 0<students 1> students.sort 2> sort.error`

表示在执行sort命令时，将一个名为students的文件中的各行进行排序，并且将排序好的内容输出到students.sort。如果由于students文件不存在而造成sort命令不能执行，错误信息被送到显示器，而不是文件sort.error。原因在于当Shell判断文件students不存在的时候，标准错误依然关联到控制台上。

例如：

`[root@localhost root]# sort 2> sort.error 0<students 1> students.sort`

表示在执行命令时，如果文件students不存在的话，错误信息将送到文件sort.error。原因在于错误重定向已经在shell判断文件students不存在之前完成了。

2.管道操作符"|"

可以用管道操作符"|"来连接各个进程，管道连接的进程可同时进行，并且能够将某个命令的标准输出作为另一命令的标准输入，例如想要将ls -l 命令所显示的内容分屏输出，则可使用一下命令：ls -l|more。

Command 1|command 2|command 3，则Command 1的输出成为command 2的输入，command 2的输出成为command 3的输入，以此类推。

例如：

`[root@localhost root]# ls -l * | wc -l`

以上命令将统计当前目录下的非隐藏文件的数目。

例如：

`[root@localhost root]#cat *.txt | wc -c`

以上命令统计当前目录下 txt 文件的总字节数。

例如：

`[root@localhost root]#ps | sort > pssort.out`

表示将 ps 命令的标准输出结果输入给 sort，经过排序后结果被保存到 pssort.out 中。

1.1.4 单引号与双引号的作用

脚本文件中的参数以用空白字符分隔，如果想在一个参数上包含一个或者多个空白字符，则需加引号。

用双引号可以引用除了$，'，\之外的字符或字符串，双引号内依然允许$的变量替换；
用单引号括起来则不允许变量替换，屏蔽几乎所有特殊含义，全部视为字符串；
用反斜杠可以消除字符的特殊含义，使字符按字面理解，包括了$。

注意：

1." "（双引号），可以防止字符串被分割,允许使用 $ 对变量进行替换。如果要在双引

号中使用 $、\、'、" 字符，则需要使用反斜线转义，如\$、\\、\'、\"。

2.'（单引号），和双引号类似，不同的是，shell 会忽略除单引号之外的所有特殊字符。因为反斜线转义在单引号中被忽略了，所以不能在单引号中使用单引号字符。

3.\（反斜线），可用来去除某些字符的特殊含义。另外，反斜线也可以实现转义的功能。

例如：
```
[root@localhost root]# A=B\\C
[root@localhost root]# echo "$A"
B\C
[root@localhost root]# echo '$A'
$A
```

例如：
```
#!/bin/sh
myvar="Hello"
echo $myvar
echo "$myvar"
echo '$myvar'
echo \$myvar
echo world
read myvar
echo '$myvar' now equals $myvar
exit 0
```

程序结果：
```
Hello
Hello
$myvar
$myvar
world
liujh      //此行为输入文本
$myvar now equals liujh
```

1.1.5 文本编辑命令 vi 或 vim 的应用

在终端输入 vi 或 vim 命令，即进入 vi 的命令模式。vi 有三种模式，分别是命令模式（command mode），插入模式（Insert mode），最后行模式（last line mode）。打开时，默认的模式为命令模式，在这个模式下可以输入许多命令。在命令模式中通过输入 a、i、A、I、o、O 等命令即可进行文本编辑。

命令模式中的指令

命令	功能
i	从光标当前位置开始输入文件
a	从目前光标所在位置的下一个位置开始追加输入文字
o	插入新的一行，从行首开始输入文字
I	在光标所在行的行首插入
A	在光标所在行的行末插入
O	在光标所在的行的下面插入一行
s	删除光标后的一个字符，然后进入插入模式
S	删除光标所在的行，然后进入插入模式
q、:q!或:qw!	退出回到终端
nG	跳到第 n 行
/单词	搜索此单词
?单词	

1.1.6 Linux 常用系统变量

环境变量在/etc/profile 文件中初始化，用来定制 shell 的运行环境，保证 shell 命令的正确执行。

    $0   当前程序的名称

    $1~$9   命令行参数 1~9 的值

    $n   当前程序的第 n 个参数，n=1, 2, …, 9

    $*   当前程序的所有参数(不包括程序本身)

    $#   当前程序的参数个数(不包括程序本身)

    $$ 当前程序的 PID

    $@   与$#相同，但是使用时加引号，并在引号中返回每个参数

    $-   显示 Shell 使用的当前选项，与 set 命令功能相同

    $!   上一个指令的 PID

    $?   上一个指令的返回值

    $HOME   当前用户主目录

    $PATH   表示当前用户的命令搜索路径，即用户不指定全路径名执行命令时，Shell 程序将在哪些目录以及按照何种顺序进行命令的搜索

    $PWD   表示用户当前所在的目录，值与 pwd 命令的结果一致

    $SHEL   表示当前用户的登录 Shell

    $USER   表示当前用户的登录名称，值与 whoami 命令的结果一致

    $UID   表示当前用户的用户名，该变量的值与 "id - u"命令的结果一致

    $LOGNAME   是指当前用户的登录名

$HOSTNAME   是指主机的名称

例如：

（1）使用 echo 命令查看单个环境变量。

`[root@localhost root]# echo $PATH`

（2）使用 env 查看所有环境变量。

`[root@localhost root]# env`

（3）使用 set 查看所有本地定义的环境变量。

（4）使用 unset 可以删除指定的环境变量。

（5）使用命令 export 设置变量，在 Shell 的命令行下直接使用[export 变量名=变量值]定义变量，该变量只在当前的 Shell(Bash)或其子 Shell(Bash)下是有效的，Shell 关闭了，变量也就失效了。

（6）在/etc/profile 文件中添加变量，该变量将会对 Linux 下所有用户有效，并且是"永久的"。

第 2 章
Shell 编程

2.1 知识要点

2.1.1 Shell 程序运行方式

假定 shell 脚本名为 filename,执行此 shell 脚本方式有:

1. 先给脚本加上可执行的权限 chmod +x filename,然后输入./filename(前提要在文件的目录下)
2. 把脚本文件作为一个参数传递给 shell: /bin/bash filename
3. 把/bin/bash 的这个目录添加到整个环境变量中:

export　PATH=/bin/bash:$PATH,然后直接输入文件名,就会执行。这个方法的优点在于可以在任何目录下编译和执行 Shell 脚本。

2.1.2 Shell 中变量的使用

注意:
1. Shell 程序中的赋值变量与等号、等号与值之间是不允许有空格的,否则就会产生语法错误。

2.变量的替换命令$(变量名)

例如：

[root@localhost root]# cmd=$(pwd)

[root@localhost root]# echo "THE VALUE OF COMMAND IS : $cmd"

结果为：

"THE VALUE OF COMMAND IS : /root"

3.变量的终端读入命令 read 变量名

让用户输入内容，并将内容储存至变量中

用法："read 变量名"

参数：

-p 后面可以跟提示符

-t 最长等待用户输入的时间

例如：

提示用户输入姓名并储存至 username 变量中

read -p "请输入您的姓名" username

2.1.3 Shell 编程中参数替换

Shell 中间提供了一些简便的参数替换的手段，这些常用于处理字符串等。

例如：

    ${param:-default}   如果 param 是空，就将它设定为 default；

    ${#param}   给出 param 的长度；

    ${param%word}   从 param 的尾部开始，删除匹配 word 的最小部分；

    ${param%%word}   从 param 的尾部开始，删除匹配 word 的最大部分；

    ${param#word}   从 param 的首部开始，删除匹配 word 的最小部分；

    ${param##word}   从 param 的首部开始，删除匹配 word 的最大部分。

以上替换通常应用在文件名和文件路径的处理上，例如常用以下代码来获取当前目录。

{PWD%/*}

2.1.4 Shell 编程中大段文字显示技巧

1.显示文件内容

从键盘读入一个文件名并显示文件内容的语句：

    read d

    cat $d.txt

2.大块文字信息显示

在 Shell 程序设计中，如果要提供大块的文字提示信息，最好使用语句块：

```
    cat << HELP
        …
    HELP
```
例如：
```
    cat << HELP
        这是一个记日记的程序
        用法：
        选项：
        -h  帮助
        -n  新日记
            进入后自动加载日期时间，安装w3m并联网后可自动写入天气。
            然后直接输入日记即可，换行后输入":wq"回车可结束输入。
        -l  显示日记
            进入后直接根据选项选择即可
        其他：
        天气的地点默认为杭州。
    HELP
```

2.1.5 随机数产生

在 Shell 中可以应用以下语句产生一个随机数字，不过这里并没有设定种子。
```
numrand=`expr $RANDOM % 999`
```

2.1.6 日期输出格式设置

在Shell程序设计中，日期命令data有丰富的输出格式，含义与终端使用格式相同。
例如：
```
%a %A %b %B %c %C %d %D %e %F %g %G %h %H,
year=`date '+%Y'`
month=`date '+%B'`
day=`date '+%d'`
echo $year
echo $month
echo $day
```
输出为：
2012
三月

2.1.7 在 Shell 中的算术表达方法

在 Shell 中算术表达可以有 let、shell 扩展 $((expression)) 和 expr 三种方法。
例如：
在 shell 中有以下几种方法表示表达式 c=a+b
```
let c=a+b
c=$((a+b))
c=$(($a+$b))
c=`expr $a + $b`
```
注意：
在此用法中 $a 与 + 之间有空格，此外表示两数相乘时应用 "*" 而不是 "*"。
```
c=$(expr $a + $b)
```
expr 命令执行速度慢，它需要调用一个新的 shell 来处理 expr 命令，更好的办法是使用 $((...)) 扩展计算表达式的值。

注意：
$((...)) 与 $(...) 不同，两对圆括号用于算术替换，而一对圆括号用于命令的执行和获取输出。

1. 用 let 运算

用 let 来计算算术表达式的值，较为简单，也可以用来变量声明，如：
```
let y=9 x=10
let z=x+y
echo $z
19
```

2. 用 $ 运算

"$" 表示取表达式或者变量的值，相当常用且方便。如变量 age=20，那么 $((100-age)) 就是 80。
例如：
```
x=$(($x+1))
```
等同于：
```
x=$(expr $x+1)
```
例如：
while [$t < 10] 也可以写成 while ((t < 10))

3. 用 expr 运算

expr 将它的参数作为表达式来求值，如 x=1，那么 $(expr $x+1) 就是 2，这只是简单应用，expr 命令可以完成许多表达式的求值运算，如 | &=<>/%*+-!= 等等，这些符号在 C 语言中已经见过了，意义也是一样的。

注意：

（1）表达式中的每个项需用空格隔开；

（2）Shell中的元字符必须使用反斜杠 \ 转义，如乘号*；

（3）对包含空格和其他特殊字符的字符串要用引号括起来。

例如：

（1）计算字串长度

```
[root@localhost root]# expr length "浙江大学"
8
```

（2）抓取字串

```
[root@localhost root]# expr substr "Zhejiang University" 6 8
ang Univ
```

（3）抓取第一个字符数字串出现的位置

```
[root@localhost root]# expr index "Zhejiang University" j
4
```

（4）整数运算

```
[root@localhost root]# expr 14 % 9
5
```

（5）增量计数

```
[root@localhost root]# LOOP=0
[root@localhost root]# LOOP=`expr $LOOP + 1`
```

expr在循环中用于增量计算。先将变量初始化为0，然后循环值加1。

（6）数值测试

```
[root@localhost root]# rr=3.4
[root@localhost root]# expr $rr + 1
```

注意：

` 与 $() 在 Shell 程序的数学运算中的区别：

例如：

x=`expr $x +1` 与 x=$(expr $x +1)一样都是把 expr $x +1 的值赋给 x，但$()可以嵌套使用，而`嵌套使用会产生歧义。

注意：

表达式 $((expression))用来对括号内的 expression 进行算术赋值且不需要在 expression 中的变量名称前加$。

例如：

```
#! /bin/bash
foo=1
foo=$((foo+1))
echo $foo
```

例如：

```
[root@localhost root]# num=$(($(wc -l<letter.txt)/22+1))
```

17

```
[root@localhost root]# echo $num
```

2.1.8 Shell 程序设计中分支语句

1.if 语句

格式为：
```
 if  [ 表达式 ]
   then 命令
 elif  [ 表达式 ]
   then 命令
 ......
 else
 命令
 fi
```
注意表达式的写法，一定要在操作数、操作符以及括号的前后至少留一个空。

例如：

输入一个数，与0比较大小
```
#! /bin/bash
read x
if [ $x -gt 0 ]
  then echo greater than 0
elif [ $x -eq 0 ]
  then echo equal to 0
else
  echo less than 0
fi
```

2.case 语句

格式为：
```
 case $变量名称 in
   "第一个变量内容")
       程序段;
       ;;
   "第二个变量内容")
       程序段;
       ;;
   *) #这里是通用匹配符
       程序段;
       ;;
```

```
  esac
```
例如：
判断传递给脚本的第一个参数，判断用户输入的 apple、banana、cat、dog
```
case $1 in
  "apple")
    echo "You choose apple"
;;
  "banana")
    echo "You choose banana"
;;
  "cat")
    echo "You choose cat"
;;
  "dog")
    echo "You choose dog"
;;
  *)
    echo "Bad param"
;;
  esac
```

2.1.9 Shell 程序设计中循环

1. for 语句

格式为：
```
for 变量名 in 所赋的值    （各个值用空格隔开）
  do
  命令
  done
```
例如：
判定学生成绩是否及格，大于等于60分视为及格。
```
#! /bin/bash
for score in 35 90 55 60 100 83 48 78
  do
    if [ $score -ge 60 ]
      then echo "$score ----- PASS"
    else
      echo "$score ----- FAIL"
```

```
    fi
  done
exit 0
```

例如：
遍历当前目录下把符合条件f*.sh的文件显示在屏幕上
```
for file in $( ls f*.sh ); do
  cat $file
done
```

2.while 语句

格式为：
```
while 条件判断
   do
     命令
   done
```

3.until 语句

格式为：
```
until 条件判断
   do
     命令
   done
```
条件判断为假时，循环执行命令，与 while 类似，但条件正好相反。

2.1.10 Shell 中函数的格式

在 Shell 程序设计中，函数的格式为：
```
function_name()
{
command_list
}
```
假如 Shell 函数中的局部变量与全局变量同名，在函数内局部变量优先，在函数外局部变量无效。函数调用时的参数传递可用$1、$2、$3…$i 来进行。

例如：
```
#!/bin/bash
yes_or_no( )
{
  echo "Is your name $*?"
  while true
```

```
      do
        echo "Enter yes or no"
        read x
        case "$x" in
        y | yes ) return 0;;
        n | no ) return 1;;
        * ) echo "Answer yes or no " ;;
        esac
      done
}

if [ $# = 0 ]
  then
     echo "Usage:myname name"
  else
     echo "Original parameters are $*"
    if yes_or_no "$1"
      then
        echo "Hi $1, nice name"
      else
        echo "Never mind"
    fi
fi
exit 0
```

假定文件名取名为kk，程序运行结果如下：

```
[root@localhost root]# ./kk   liujiahai
Original parameters are liujiahai
Is your name liujiahai?
Enter yes or no
y
Hi liujiahai, nice name
[root@localhost root]#
```

2.1.11 Shell 中数组的使用

数组是存储在连续内存空间的相同类型的一组元素。数组的下标是整数并以数字 0 作为起始，即数组的第 1 个元素的下标为 0，通过数组的索引和下标去引用。

1. 创建数组

    name=(value1 value2 ...[下标1]=字符串1 [下标2]=字符串2 ....)

2. 引用元素

    ${name[下标]}

3. 引用数组

    ${name[@]}或${name[*]}

    前者原样复制，后者将原数组所有元素作为新数组的第一个元素。

4. 数组初始化

例如：

 names[2]=AAA   将AAA定义为数组names的第3个元素

 names[0]=BBB   将BBB定义为数组names的第1个元素

也可以多个同时初始化：用一个小括号括起来！

 names=([2]=AAA [0]=BBB [1]=CCC)

也可以利用循环脚本初始化，这种方法适用于元素较多的数组。

值得关注的是，对于排序的中间缺值的，如果想跳过直接赋值后面的，可使用：

names=(AAA [5]=BBB CCC)

则数组的names[0]是AAA，names[5]是BBB，names[7]是CCC

如果用*作为下标，则显示数组的元素个数。

例如：

分析下列程序段，其功能是给数组a[1]-a[10]赋值，请思考运算符le、数组等应用。

```
i=1
n=1000
while $i -le 10
do
        a[$i]=$n%2
        n=$n/2
        i=$i+1
done
```

 ## 2.1.12 文件与文件夹的判断

在 Linux 中，文件的判断用参数-f，文件夹的判断用参数-d 进行。

例如：

从键盘输入字符串到 Dir，然后用参数-d 判断 Dir 是否是文件夹

```
read Dir
if [ -d $Dir ]
```

例如：

在 Shell 程序设计中，判断正在使用的 Linux 系统是否安装有某个软件，如 CD 播放器，

如有安装，执行 CD 播放器程序。程序代码可以写为：
```
cd /usr/bin
if [ -f   gnome-cd ]          #查看是否安装了CD播放器
   then
   gnome-cd                   #运行CD播放器
else
   echo "Your system don't have the CD Player!"
   echo "Press any key to return......"
fi
```

2.1.13 某路径下文件总容量的判断

获取某个路径下文件的总存储容量
例如：
要测试/var/mail/09207041 路径下文件的总存储容量可以使用表达式：
```
count1=`ls -l /var/mail/09207041|awk '{print $5}'`
```

2.1.14 菜单程序的框架

在 Shell 程序设计中，往往采用以下框架：
```
while:
do
echo "1. ....... "
echo "2. ....... "
echo "3. ....... "
echo "4. ....... "
echo "q. quit"
read CHOICE
case $CHOICE in
 1)
 .......
 ;;
 2)
   .......
   ;;
 3) .......
   ;;
```

```
    4) .......
      ;;
    q)exit 1;;
esac
done
```

2.2 程序设计实例

实例1 应用菜单、循环、分支等 Shell 语法结构，编写的文件压缩和解压缩程序界面并实现其功能。

```
#!/bin/sh
#Linux压缩&解压缩工具
#免除记忆命令的烦恼
compression()
{
if [ ! -d "$DESTINATION" -o ! -f "$DESTINATION"];
then
      echo "该路径不存在！"
      return
fi
#if [ ! -x "$DESTINATION" ]
#then
      #echo "对不起，您没有操作该路径的权限。"
      #return
#fi
clear
echo "================================="
echo "==        请选择压缩格式          =="
echo "==        1-*.bz2                =="
echo "==        2-*.gz                 =="
echo "==        3-*.tar                =="
echo "==        4-*.tar.gz             =="
echo "==        5-*.zip                =="
echo "================================="
read COMCHOICE
```

```
#echo $COMCHOICE
case $COMCHOICE in
    1) bzip2 -z $DESTINATION;;
    2) gzip -r $DESTINATION;;
    3) tar -cvf $DESTINATION.tar $DESTINATION;;
    4) tar -zcvf $DESTINATION.tar.gz $DESTINATION;;
    5) zip -r $DESTINATION.zip $DESTINATION;;
    *) echo "输入有误！正确的输入为（1--5）";;
esac
}

decompression()
{
if [ ! -f "$DESTINATION" ]
then
    echo "该文件不存在！"
    return
fi
#if [ ! -x "$DESTINATION" ]
#then
    #echo "对不起，您没有操作该文件的权限。"
    #return
#fi
clear
echo "================================="
echo "==         请选择解压缩的格式         =="
echo "==            1-*.bz2            =="
echo "==            2-*.gz             =="
echo "==            3-*.tar            =="
echo "==            4-*.tar.gz         =="
echo "==            5-*.zip            =="
echo "================================="
read DECCHOICE
#read $DECCHOICE
case $DECCHOICE in
    1) bzip2 -d $DESTINATION;;
    2) gzip -d $DESTINATION;;
    3) tar -xvf    $DESTINATION;;
    4) tar -zxvf   $DETINATION;;
```

```
     5) unzip $DESTINATION;;
     *) echo "输入有误！正确的输入为（1--5）";;
esac
}

quit()
{
clear
echo "================================"
echo "==      Linux压缩与解压工具      =="
echo "==                              =="
echo "==          谢谢使用！           =="
echo "==                              =="
echo "================================"
exit 0
}

clear
while true
do
    echo "================================"
    echo "==      Linux压缩与解压工具      =="
    echo "==           1-压缩             =="
    echo "==           2-解压             =="
    echo "==           3-离开             =="
    echo "================================"
    echo "请输入选择(1-3)"
    read CHOICE
    #echo $CHOICE
    case $CHOICE in
        1) echo "请输入所要压缩的文件或文件夹"
           read DESTINATION
           echo $DESTINATION
           compression $DESTINATION;;
        2) echo "请输入所要解压的文件或文件夹"
           read DESTINATION
           echo $DESTINATION
           decompression $DESTINATION;;
        3) quit;;
```

```
        *) echo "输入有误！正确的输入为（1--3）"
           sleep 3
           clear
      esac
   done
```

实例2 监控CPU和内存。显示系统的日期和时间，每隔五分钟将CPU和内存在相应时刻的使用情况信息输出到文件capstats.csv，便于监控CPU和内存的使用情况，假定程序名为2-2。

```
#!/bin/bash
#输出文件为capstats.csv
OUTFILE=/root/capstats.csv
#显示当前日期
DATE='date +%Y%m%d'
#显示当前时间
TIME='date +%k:%M:%S'
TIMEOUT='uptime'
#显示用户数量
USERS='echo $TIMEOUT | gawk '{print $4}''
#显示系统5分钟内的平均负载
LOAD='echo $TIMEOUT | gawk '{print $9}' | sed 's/,//''
#显示系统未使用的物理内存量
FREE='vmstat 1 2 | sed -n '4p' | gawk '{print $4}''
#系统CPU空间的时间百分比
IDLE='vmstat 1 2 | sed -n '4p' | gawk '{print $15}''
#将各项信息输入到文件capstats.csv
echo "date: $DATE, time:$TIME, users:$USERS, load:$LOAD, free:$FREE, idle:$IDLE">>$OUTFILE
```

（2）设置权限：

```
[root@localhost root]# chmod +x 2-2
```

（3）每隔5分钟执行一次：

①`[root@localhost root]# crontab -e`

②在vi中编辑：`*/5 * * * * /root/2`

③保存并退出，程序将会自动每隔5分钟执行一次

（4）查看运行结果

```
[root@localhost root]# cat capstats.csv
```

实例3 程序实现了用户交互对话模式，对ROOT用户实现自动检测功能，可以对本账户下的所有进程加以体现与控制。

```
#!/bin/sh
menu(){   #功能选择界面menu菜单
```

```bash
  clear
  echo "******************************"
  echo "*    Please choose one function"
  echo "*  1) Date information"
  echo "*  2) Create a new user"
  echo "*  3) Delete an existing User"
  echo "*  4) Show all running programs"
  echo "*  5) Quit"
  echo "******************************"
}

dateinfo(){     #时间信息显示函数
  clear
  echo "******************************"
  echo "*     Mr.$USER,Today is:"
  echo  &date "+%B%d%A"
  echo "*     Wish you a lucky day !"
}

usercreate(){   #用户账户创建函数
  clear
  echo "******************************"
  echo "Create a new user"
  echo "--------------------------------"
  echo "Enter a name for the new user: "
  read USERNAME

  useradd $USERNAME
  if(($? == 0))              #判断是否创建成功
  then
    echo "--------------------------------"
    echo "You've created $USERNAME !"
  else
    echo "--------------------------------"
    echo "create error !"
  fi
}

userdelete(){          #用户账户删除函数
```

```
    clear
    echo "*****************************"
    echo "*You are going to delete an existing user!"
    echo "--------------------------------"
    echo "Enter the name of an existing user: "
    read USERNAME

    userdel $USERNAME
    if(($? == 0))              #判断是否删除成功
    then
      echo "--------------------------------"
      echo "You've deleted $USERNAME !"
    else
      echo "--------------------------------"
      echo "delete error !"
    fi
}

listprocess(){              #列出指定账户下的进程
    clear
    echo "*****************************"
    echo "*  You are going to show running programs"
    echo "--------------------------------"
    echo "Enter the name of an existing user: "
    read USERNAME

    clear
    echo "Running programs under [$USERNAME]: "
    echo "--------------------------------"
    ps -u $USERNAME
}

quit(){    #退出菜单显示功能
    clear
    echo "*********************************"
    echo "*           Thank you for your use!"
    echo "*     The program will exit in 5 seconds!"
    echo "*********************************"
    sleep 5
```

```
}


#main主函数
echo "*******************************"
echo "*"
echo "*              Welcome $USER!"
echo "*   The program will run in 5 seconds !"
echo "*******************************"
sleep 5
clear

if test $USER = "root"          #判断是否为ROOT账户
then
  userflag=0

    while(($userflag == 0)) #总循环，账户为ROOT账户
      do
        menu
        userchoice=0          #初始化  用户选择默认为0
          while (($userchoice == 0))
    do
      echo "Please enter your choice in (1-5):"
      read userchoice
      case $userchoice in
         1) dateinfo;;
         2) usercreate;;
         3) userdelete;;
         4) listprocess;;
         5) userflag=1;;
         *) echo "Invalid input. A correct choice should be in the range 1-5."
            userchoice=0;;      #用户功能选择出错，提示返回
    esac
    done
       if((userflag == 0))
       then
         echo "*******************************"
         echo "*     Press ENTER to return!"
```

```
            read
        fi
    done
    quit
    clear   #退出清屏
else
  userflag=0
  echo "You've logged in as a user with low authority;"
  echo "Please log in as [root] user! THX!"
fi
```

第 3 章 Linux 系统 C 语言开发工具

3.1 知识要点

3.1.1 gcc 的使用

在 Linux 下，gcc 较常用的一种格式为：

gcc 源文件名 -o 目标文件名

例如：

与 Windows 系统不同，目标文件名不一定非要以.exe 为后缀，但是源文件名应该以.c 结尾。例如编辑源程序文件 main.c。

```
#include <stdio.h>
int main(int argc, char* argv[])
{
int i = 0;
for(i = 0; i < argc; i++)
    printf("Argument %d : %s\n", i, argv[i]);
return 0;
}
```

这个程序打印出了所有接受到的参数。下面使用 gcc 来编译它：

gcc main.c -o main

执行之：

./main

注意：

#include<>表示在默认路径 "/usr/include" 中搜索头文件

而#include " " 表示在本目录下搜索头文件

在用 gcc 编译时要用到-l 或-L 参数。比如在用 math.h 库时，需要加上-lm。

放在/lib 和/usr/lib 和/usr/local/lib 里的库直接用-l 参数就能链接了，但如果库文件没放在这三个目录里，而是放在其他目录里，这时需要使用参数-L 编译。比如常用的 X11 的库，它在/usr/X11R6/lib 目录下，在编译时就要用-L/usr/X11R6/lib -lX11 参数，-L 参数跟着的是库文件所在的目录名。

例如：

把 libtest.so 放在/aaa/bbb/ccc 目录下，那链接参数就是

-L/aaa/bbb/ccc -ltest

3.1.2 makefile 工程文件

1.makefile 文件的书写规则

目标文件：依赖文件

（tab）产生目标文件的命令

变量的使用应注意：

（1）变量名不能有":"、"#"、"="或是空字符（空格、回车等）；

大小写敏感，一般都用大写；

在声明时需赋初值，使用时要在前面加 "$" 符号；

2.makefile 文件的编写

makefile 文件的编写可以说是作为一个程序员必备的技能，特别是在 Unix/Linux 环境下工作的程序员们，而 makefile 的好处就是可以实现 "自动化编译"，编写了 makefile 文件之后，只需要一个 make 命令，就能方便地编译整个工程。

对于大型的工程来说，makefile 中的变量就显得犹为重要。在此论述一下常用的 makefile 变量。

（1）用 OBJECT、OBJS 来表示最终目标的文件列表

例如：

OBJS=main.o f1.o f2.o f3.o

可以使用$来引用，这样最终目标的编译和最后的 clean 可以表示为：

program: ${OBJS}

    gcc -o program ${OBJS}

    clean:

```
rm ${OBJS}
```

（2）使用预定义的变量

$@ 表示当前目标文件的名字。

$^ 表示用空格隔开的所有依赖文件。

那么上述的编译过程可以简化为：

```
program: ${OBJS}
    gcc -o $@ $^
```

（3）让 makefile 自动推导，make 命令可以自动推导文件以及文件依赖关系后面的命令，make 会自动识别并自己推导命令。只要 make 查到某个.o 文件，它就会自动的把此.c 文件加到依赖关系中。例如：如果 make 找到一个 whatever.o，那么 whatever.c 就确定为是 whatever.o 的依赖文件。并且 gcc -c whatever.c 也会被推导出来，于是 makefile 也不用写得这么复杂。假设 f1、f2、f3 分别包含了 math.h、stdlib.h 和 ctype.h，则：

OBJS=main.o f1.o f2.o f3.o

program: ${OBJS}

    gcc –o $@ $^

例如：

makefile 文件的形式为：

```
main:main.o mytool1.o mytool2.o
    gcc -o main main.o mytool1.o mytool2.o
main.o:main.c mytool1.h mytool2.h
    gcc -c main.c
mytool1.o:mytool1.c mytool1.h
    gcc -c mytool1.c
mytool2.o:mytool2.c mytool2.h
    gcc -c mytool2.c
```

应用$@目标文件、$^所有的依赖文件、$<第一个依赖文件把上述 makefile 的简化：

```
main:main.o mytool1.o mytool2.o
    gcc -o $@ $^
main.o:main.c mytool1.h mytool2.h
    gcc -c $<
mytool1.o:mytool1.c mytool1.h
    gcc -c $<
mytool2.o:mytool2.c mytool2.h
    gcc -c $<
```

应用缺省变量，把上述 makefile 进一步简化为：

```
main:main.o mytool1.o mytool2.o
    gcc -o $@ $^
..c.o:
    gcc -c $<
```

3.2 程序设计实例

实例1 设计一个程序,从键盘输入的数字到变量 a,要求算出它的 sin(a)值。程序中首先从键盘输入一个数,然后用循环的方法输入实型数,用函数 sin 计算后输出,本题关键的问题是学习如何编译。

1.源程序代码

```
[root@localhost root]#vim  3-1.c
#include<stdio.h>            /*文件预处理,包含标准输入输出库*/
#include<math.h>             /*文件预处理,包含数学函数库*/
int main()                   /*C 程序的主函数,开始入口*/
{
    double a,b;
    printf("请输入自变量:");
    scanf("%lf",&a);
    b=sin(a);                /*调用数学函数计算*/
    printf("sin(%lf)=%lf\n",a,b);
}
```

2.用 gcc 编译程序

```
[root@localhost root]#gcc  3-1.c  -o  3-1
```

结果发现编译器报错,具体的提示如下:

```
/tmp/ccjPJnA.o(.text+0x3e): In function 'main':
: undefined reference to 'sin'
collect2: ld returned 1 exit status
```

虽然包含数学函数库,但还是提示没有定义函数 sin,原因是还需要指定函数的具体路径,这首先要对函数进行查找。函数的查找方法如下:

```
[root@localhost root]#nm -o /lib/*.so|grep 函数名
```

例如:

要查找函数 sin,在终端输入的命令如下:

```
[root@localhost root]# nm -o /lib/*.so|grep  sin
```

这时查找的结果中有部分内容显示如下:

```
............................
/lib/libm-2.3.2.so:00008610 W sin
/lib/libm-2.3.2.so:00008610 t __sin
............................
```

在/lib/libm-2.3.2.so:00008610 W sin 中,除去函数库头 lib 及函数的版本号-2.3.2,所余

下的符号为"m",在编译时用字符"1"与余下的符号"m"相连接成"lm",在编译时加上此参数,即:

[root@localhost root]# **gcc 3-1.c -o 3-1 -lm**

就能正确地通过编译。

3.运行程序

编译成功后,执行可执行文件 3-1。

实例 2 程序在路径/home/exp/basic/有一库文件 liujh.h,功能是调用函数 funmax 求出数组的最大值,在路径/root 有主文件 test3-2.c,在此文件中输入数组元素,并对元素开平方。

1.Test3-2.c 源程序

```
[root@localhost root]# vi test3-2.c
#include "liujh.h"
int main()
{
  int a[10],i,max;
  for(i=0;i<10;i++)
    scanf("%d",&a[i]);
  for(i=0;i<10;i++){
    if(a[i]<0)
      a[i]=fabs(a[i]);
    a[i]=(int )fqurt(a[i]);
    }
  max=funmax(a,10);
  printf("max=%d\n",max);
  return 0;
 }
```

2.在路径/home/exp/basic/liujh.h 源代码

```
#include<stdio.h>
int funmax(int a[],int n)
{
  int i,max;
  max=a[0];
  for(i=1;i<n;i++)
    if(max<a[i])
       max=a[i];
   return max;
}
```

3.编译

[root@localhost root]# gcc -o test3-2 test3-2.c -I /home/exp/basic -lm

第 4 章
Linux 环境下系统函数的使用

4.1 知识要点

4.1.1 随机函数的应用

要使在每次运行时产生不同的随机数,首先要使用随机种子函数 srand,一般来说使用每天的时间 srand((unsigned int)time(0));作为随机数产生器的种子,例如下列表达式能产生 1~100 间的随机数。

```
        srand((unsigned)time(NULL));        //产生随机数
        test = rand()%100 + 1;
```

利用 srand()函数设置随机数种子,表达式 1+(int)(10.0*rand()/RAND_MAX+1.0)可以产生 1~10 之间的一个随机数。如需要 1~50 之间的随机数,可用表达式:

```
        rand()%50+1
```

要产生从 x1 到 x2 的随机数可以写成:

```
        k=rand()%(x2-x1+1)+x1;
```

例如:
在屏幕上随机显示一个英文字母

```
char str[] = "abcdefghijklmnopqrstuvwxyz";
len = strlen(str);
```

```
srand((int)time(0));
k = len * 1.0f * rand() / RAND_MAX;
printf("%d: %c --- ", (int)k+1, str[(int)k]);
```

4.1.2 系统时间函数的应用

常用时间函数有 asctime、time、gmtime、ctime、localtime、gettimeofday，时间结构体 struct tm 如下：

```
struct tm
{
    int tm_sec;        //代表目前秒数，0~59
    int tm_min;        //代表目前分数，0~59
    int tm_hour;       //
    int tm_mday;
    int tm_mon;
    int tm_year;
    int tm_day;
    int tm_yday;
    int tm_isdst;
};
```

tm_year 为 1900 算起至今的年数，因此在计算现在年份时加 1900，tm_mon 范围为 0~11，而真实月份为 1~12，所以计算月份时加 1，年中天数和礼拜中星期几的计算同理。

其中两个结构体如下：

```
struct timeval
  {
      time_t      tv_sec;       /* seconds */
      suseconds_t tv_usec;      /* microseconds */
  };
struct timezone
 {
    int tz_minuteswest;      /* minutes west of Greenwich */
    int tz_dsttime;          /* type of DST correction */
 };
```

这个函数比较特殊，利用它就可以很方便地得到目前的时间，并且精确到微秒，通常可以利用它的高精度来计时。

例如：

```
time_t t;             //创建时间类型变量 t
time(&t);             //将从 1970.1.1 零点起到现在经过的秒数存入 t
```

```
printf("%s", asctime(localtime(&t)));    //按格式输出表示当前时间的字符串
```
例如：
以 ASCII 码形式输出本地时间语句为：
```
t=time(0);
printf("当前系统的时间为： %s\n",asctime(localtime(&t)));
```
例如：
定义结构体 timeval 变量，分别存储某事件前与后的系统时间，则事件前、后的总时间可用程序段表示：
```
struct timeval tv1,tv2;
sec = tv2.tv_sec - tv1.tv_sec;
usec = (tv2.tv_usec - tv1.tv_usec) * 1e-6;    //1秒=10^6 微秒
sum = sec + usec;
```

4.1.3 系统函数

1. 系统函数 mkdir

此函数在建立路径的同时，可以设置路径的权限，例如：
```
mkdir("dir",S_IXUSR|S_IXGRP|S_IXOTH)
```

2. 系统函数 system

system()会调用 fork()产生子进程，由子进程来调用/bin/sh-c string 来执行参数 string 字符串所代表的命令，此命>令执行完后随即返回原调用的进程。在调用 system()期间 SIGCHLD 信号会被暂时搁置，SIGINT 和 SIGQUIT 信号则会被忽略。

例如：
```
system("cat -n /var/spool/mail/root > /root/mailtest.log");
```
例如：
应用 system 函数，使用命令在固定的时间关闭计算机。
```
#include <time.h>
#include <stdlib.h>
#include <unistd.h>
#define DELAY 60                    /* 睡眠的时间 */

int main()
{
  time_t now;
  struct tm *p;
  while(1)
  {
    now = time(NULL);
```

```
    sleep(DELAY);                    /*减少cpu的占用*/
    p = localtime(&now);
    if ((p->tm_hour == 14) && (p->tm_min >= 30))
     /*系统将在14:30或14:31关闭*/
    system("poweroff");               /*'poweroff'是指linux系统将关闭*/
    }
    return 0;
}
```

4.1.4 数据结构中的函数

函数 qsort、bsearch、lfind 主要功能是二分查找、快速排序、线性搜索。
1.快排函数 qsort
`qsort(num,n,sizeof(int),compar);`
num 是待排序的数组（这里为 int 型），n 为数组元素个数，函数 compar 决定排序规则。
注意：

在数据排序时用到函数 compar

```
int compar(const void *a, const void *b)
{
    int *aa = (int *)a, *bb = (int *)b;
    if (*aa > *bb) return 1;
    if (*aa == *bb) return 0;
    if (*aa < *bb) return -1;
}
```

2.二分查找的函数 bsearch
`Find=(int *)bsearch(&key,num,n,sizeof(int),compare);`
　　值得注意的是 bsearch 带有返回值 返回的是被查找的数 key 在数组中的地址，如果找不到则返回 0。与 qsort 中形参不同的是，bsearch 中还多了一个 &key 表示待查找数的地址，值得注意的是该处是地址而不是元素。
思考：
　　函数 compar 把 1 及-1 作一个交换，对程序运行结果有什么影响？是否可以对字符串"aaabc"、"aabc"、"aaaade"等进行排序？

4.2 程序设计实例

实例 1 设计一个程序,在屏幕上随机显示一个个位数,并开始等待用户键盘输入字符,计算从字符显示到用户输入所用的时间,多次重复此过程,最后输出用户的平均正确反应时间(输入错误的不计入)以及正确率。源程序:

```
#include <stdio.h>
#include <stdlib.h>
#include <time.h>
#include <string.h>
#include <sys/time.h>
#include <unistd.h>

int main()
{
    struct timeval tv1, tv2;
    struct timezone tz;
    float tm = 0.0f;
    int repeat = 25,i,crt = 0,k,n;
    for (i=0; i<repeat; i++)
    {
        /*随机产生个位整数*/
        srand((int)time(0));
        k = (int) ( 10.0 * rand() / (RAND_MAX+1.0));
        printf("%d: %d --- ", i+1, k);
        /*取得输入前后的时间来计算时间差*/
        gettimeofday(&tv1, &tz);
        scanf("%d",&n);
        gettimeofday(&tv2, &tz);
        /*如果输入正确,计算正确率和平均反应时间*/
        if (n == k)
        {
            crt ++;
            tm+=(tv2.tv_sec-tv1.tv_sec)+(tv2.tv_usec-tv1.tv_usec)/100000.0f;
        }
    }
```

```c
    tm /= crt * 1.0f;
    printf("Correctness Rate: %d%%\n", (int) (crt * 100.0 / repeat));
    printf("Average time taken: %.2f sec(s)\n", tm);
}
```

实例 2 使用系统时间函数来测试几个排序算法的效率,这几个算法分别是插入法排序(Insertion Sort),二分插入法排序(Half Insertion Sort),希尔排序(Shell Sort)。首先使用 rand 函数来随机生成一组数,由于个数定义为宏,可以通过直接修改宏来实现个数的调整。下述代码以 10000 个数为例。程序代码为:

```c
#include <stdio.h>
#include <time.h>
#include <stdlib.h>
#include <unistd.h>
#include <sys/time.h>
#define LEN 10000
#define N 3
/*Insertion Sort*/
void InsertionSort(long input[],int len)
{
    int i,j,temp;
    for (i = 1; i < len; i++)
    {
        temp = input[i];                    /* 操作当前元素,先保存在其它变量中 */
        for (j = i - 1;j>-1&&input[j] > temp ; j--)
        {
            input[j + 1] = input[j];        /* 一边找一边移动元素 */
            input[j] = temp;
        }
    }
}


/* Shell Sort */
void ShellSort(long v[],int n)
{
    int gap,i,j,temp;
    for(gap=n/2;gap>0;gap /= 2)             /*设置排序的步长,步长 gap 每次减半,直到减到 1*/
    {
        for(i=gap;i<n;i++)                  /* 定位到每一个元素 */
        {
            for(j=i-gap;(j >= 0) && (v[j] > v[j+gap]);j -= gap )
```

```c
                {
                 temp=v[j];
                 v[j]=v[j+gap];
                 v[j+gap]=temp;
                }
        }
    }
}

/* HalfInsertSort */
void HalfInsertSort(long a[], int len)
{
    int i, j,temp;
    int low, high, mid;
    for (i=1; i<len; i++)
    {
        temp = a[i];/* 保存但前元素 */
        low = 0;
        high = i-1;
        while (low <= high)     /* 在a[low...high]中折半查找有序插入的位置 */
        {
            mid = (low + high) / 2;        /* 找到中间元素 */
            if (a[mid] > temp)
          /* 如果中间元素比但前元素大, 当前元素要插入到中间元素的左侧 */
            {
             high = mid-1;
            }
            else    /* 如果中间元素比当前元素小, 但前元素要插入到中间元素的右侧 */
            {
             low = mid+1;
            }
        }        /* 找到当前元素的位置, 在low和high之间 */
        for (j=i-1; j>high; j--)        /* 元素后移 */
        {
         a[j+1] = a[j];
        }
        a[high+1] = temp;        /* 插入 */
    }
}
```

```c
double test(void (*fp)(long a[], int len))
{
    long a[LEN];
    double b;
    int i,j;
    struct timeval tv1,tv2;
    struct timezone tz;
    j = (int)time(0);
    srand(j);
    for(i=0;i<LEN;i++)
    {
        a[i]=rand();
        srand(++j);
    }
    gettimeofday(&tv1,&tz);
    (*fp)(a,LEN);
    gettimeofday(&tv2,&tz);
    b = (tv2.tv_sec-tv1.tv_sec)+(tv2.tv_usec-tv1.tv_usec)/1000000.0;
    return b;
}
int main()
{
    printf("     Insertion Sort : %8f\n",test(InsertionSort));
    printf("Half Insertion Sort : %8f\n",test(HalfInsertSort));
    printf("         Shell Sort : %8f\n",test(ShellSort));
    return 0;
}
```

实例 3 程序设计。程序运行时首先呈现文件解压缩的菜单选择。然后输入相应的解压缩文件名，最后给出解压缩是否成功的提示。

```c
/*包含必须的头文件....*/
#include<stdio.h>
#include<stdlib.h>
#include<string.h>

/*压缩文件*/
void zip(int format)
{
/*执行压缩时候的命令*/
```

```c
    char cmd1[50]="gzip ";
    char cmd2[50]="tar -cf ";
    char file[30];          /*有路径的文件名*/
    switch(format)
    {
      case 1:
        printf("请输入您要压缩的文件名(如需多个文件请以空格作为间隔): ");
        getchar();
        fgets(file,30,stdin);
        strcat(cmd1,file);
        system(cmd1);
        printf("您所需要的文件已经成功压缩!\n");
        break;
      case 2:
printf("您需要的目标文件名 然后加空格在输入您要归档的文件,(如需多个文件请以空格作为间隔)\n");
        getchar();
        fgets(file,30,stdin);
        strcat(cmd2,file);
        system(cmd2);
        printf("您所需要的文件已经成功压缩!\n");
        break;
      default :
        printf("您的选择有误,请重新运行该程序!\n");
        exit(0);
    }
}

/*解压缩文件*/
void unzip(char filename[])
{
    int len;
    char cmd1[30]="gunzip ";
    char cmd2[30]="tar -xvf ";

    len=strlen(filename);
    /*判定输入进来的文件的格式*/
    if(filename[len-1]=='z'&&filename[len-2]=='g')
    {
        strcat(cmd1,filename);
```

```c
        system(cmd1);
        printf("文件已经解压缩！\n");
    }
    else if(filename[len-1]=='r'&&filename[len-2]=='a'&&filename[len-3]=='t'){
        strcat(cmd2,filename);
        system(cmd2);
        printf("文件已经解压缩！\n");
    }

}
/*主函数从这里开始*/
int main(){
/*声明变量*/
int choice;/*选择压缩或者是解压*/
int format;/*选择压缩或格式*/
char filename[30];

/*做一个简单的界面*/
printf("_____\n");
printf("欢迎使用此压缩解压程序！\n");
printf("请输入您的选择:");
printf("1:压缩\t2:解压缩\n");
label:
scanf("%d",&choice);
getchar();
switch(choice){
    case 1:
        printf("您选择了压缩，请输入您需要的压缩格式：\n");
        printf("1: .gz\t2: .tar\n");
        scanf("%d",&format);
        getchar();
        zip(format);
        break;
    case 2:
        printf("您选择了解压缩，请输入您需要解压缩的文件：");
        getchar();
        fgets(filename,30,stdin);
        unzip(filename);
```

```
        default:
            printf("您的选择有误,请您重新输入!\n");
            goto label;

    }
    printf("_____\n");
    return 0;
}
```

第 5 章
Linux 环境下文件 I/O 操作

5.1 知识要点

5.1.1 文件操作

1.文件描述符

对 Linux 而言，文件描述符是一个索引值，对设备与文件的操作都是使用文件描述符来进行的，它指向在内核中每个进程打开文件的记录表。在文件 I/O 操作的系统调用时，主要用到的 5 个函数是 open、read、write、lseek 和 close。

低级 I/O 函数在很多应用中是不带缓冲区的并且通过内核完成操作，操作效率较低，应用于开发驱动程序。

2.获取文件属性函数 stat、fstat

函数 stat 是在文件关闭时测试文件的属性，而函数 fstat 应用在文件打开时测试属性。

int stat(const char *restrict pathname, struct stat *restrict buf);提供文件名字，获取文件对应属性。

int fstat(int filedes, struct stat *buf);通过文件描述符获取文件对应的属性。

例如：

```
#include <unsitd.h>
#inlcude <sys/stat.h>
```

```
#include <sys/types.h>
int main()
{
    struct stat *ptr;
    stat("/etc/passwd",ptr);
    printf("The major device no is:%d\n",major(ptr->st_dev));//主设备号
    printf("The minor device no is:%d\n",minor(ptr->st_dev));//从设备号
    return 0;
}
```

3.文件类型的判断

文件类型使用参数 st_mode 通过宏来判断属性。

例如：

如果 S_ISLNK(st_mode)为真，即是一个连接文件，S_ISREG(st_mode)判断是否是一个常规文件，S_ISDIR(st_mode)判断是否是一个目录文件，S_ISFIFO(st_mode)是否是一个FIFO 文件，S_ISSOCK(st_mode)是否是一个 SOCKET 文件等等。

例如：

文件属性的判断

```
void show_type(struct stat &buf)   //此函数用于显示文件的类型
{
if (S_ISLNK(buf.st_mode)) printf("It's a Link File.\n");
else if (S_ISCHR(buf.st_mode)) printf("It's a Character Device.\n");
else if (S_ISDIR(buf.st_mode)) printf("It's a Directory.\n");
else if (S_ISREG(buf.st_mode)) printf("It's a Regular File.\n");
else if (S_ISBLK(buf.st_mode)) printf("It's a Block Device.\n");
else if (S_ISFIFO(buf.st_mode)) printf("It's a FIFO File.\n");
else if (S_ISSOCK(buf.st_mode)) printf("It's a SOCKET File.\n");
else printf("It's an unknown-type file.\n");
}
```

例如：

文件权限的判断，分别判断用户、组、其他人的读、写、可执行的文件权限。

```
void show_per(struct stat &buf)   //次函数用于显示文件的权限信息
{
printf("User:");
if(S_IRUSR&(buf.st_mode))
    printf("Read   ");
if(S_IWUSR&(buf.st_mode))
    printf("Write   ");
if(S_IXUSR&(buf.st_mode))
    printf("Execute");
```

```
        printf("\nGroup:");
        if(S_IRGRP&(buf.st_mode))
            printf("Read   ");
        if(S_IWGRP&(buf.st_mode))
            printf("Write   ");
        if(S_IXGRP&(buf.st_mode))
            printf("Execute");
        printf("\nOther:");
        if(S_IROTH&(buf.st_mode))
            printf("Read   ");
        if(S_IWOTH&(buf.st_mode))
            printf("Write   ");
        if(S_IXOTH&(buf.st_mode))
            printf("Execute");
        printf("\n");
    }
```

例如：

文件属性中的用户名、所属组的判断

```
struct stat buf;
lstat(argv[1], &buf);
struct passwd *usr;
struct group *grp;
usr = getpwuid(buf.st_uid);
grp = getgrgid(buf.st_gid);
printf(" %s %s %d", usr->pw_name, grp->gr_name, buf.st_size);
```

例如：

　　应用函数 opendir 打开文件夹/var/spool/mail/owner，然后用指针 ptr 指向函数 readdir 读取此文件夹下的文件与目录不为空时，拷贝 /var/spool/mail/owner 的所有文件到 /home/owner/mail.log 中。

```
dir=opendir("/var/spool/mail/owner");
if ((ptr=readdir(dir))!=NULL)
    system("cat /var/spool/mail/owner/* > /home/owner/mail.log");
```

例如：

以下程序段用于列出目录中所有文件的名字。

```
if(dp = opendir(argv[1] == NULL)
    err_sys("can`t open %s ",argv[1]);
while((dirp = readdir(dp))!= NULL)
    printf("%s\n",dirp->d_namlen);
```

5.1.2 文件控制特性的判断

1. 死锁

比如进程 1 访问文件 A 锁死了它，进程 2 访问文件 B 锁死了它，然后进程 1 希望访问文件 B，因为进程 2 锁死了它而进入等待，进程 2 希望访问文件 A，因为进程 1 锁死了它而进入等待，如此往复，文件被锁死。

2. 文件上锁函数 flock 和 fcntl

在 Linux 中，实现文件上锁的函数有 flock 和 fcntl，其中 flock 用于对文件施加建议性锁，而 fcntl 不仅可以施加建议性锁，还可以施加强制锁。同时，fcntl 还能对文件的某一记录进行上锁，也就是记录锁。记录锁又可分为读取锁和写入锁，其中读取锁又称为共享锁，它能够使多个进程都能在文件的同一部分建立读取锁。而写入锁又称为排斥锁，在任何时刻只能有一个进程在文件的某个部分上建立写入锁。

可以用 fcntl 函数改变一个已打开文件的属性，可以重新设置读、写、追加、非阻塞等标志，使用函数 fcntl 通过 F_GETFL、F_SETFL 可以分别用于读写、设置文件的属性，能够更改的文件标志有 O_APPEND、O_ASYNC、O_DIRECT、O_NOATIME 和 O_NONBLOCK。

获取文件的 flags，即 open 函数的第二个参数；
```
flags=fcntl(fd,F_GETFL,0);
```
设置文件的 flags；
```
fcntl(fd,F_SETFL,flags);
```
增加文件的某个 flags，比如文件时阻塞的，想设置成非阻塞；
```
flags=fcntl(fd,F_GETFL,0);
flags |=O_NONBLOCK;
fcntl(fd,F_SETFL,flags);
```
取消文件的某个 flags，比如文件是非阻塞的，想设置成阻塞的；
```
flags=fcntl(fd,F_GETFL,0);
flags &=~O_NONBLOCK;
fcntl(fd,F_SETFL,flags);
```
具体文件锁的操作由函数 fntl 完成，原型是
```
int fcntl(int fildes,int cmd,struct flock *arg);
```
其中，结构 flock 用于描述文件锁的信息，定义于"fcntl.h"中，如下表示：
```
struct flock
{
    short l_type;      /*锁类型 */
    short l_whence;    /*锁区域开始地址的相对位置 */
    long l_start;      /*锁区域开始地址偏移量 */
    long l_len;        /*锁的长度，0 表示锁至文件末*/
```

```
    long l_pid;         /*拥有锁的进程 ID 号*/
};
```

锁类型：取值为 F_RDLCK 读锁、F_WRLCK 写锁、F_UNLCK 释放锁；

锁区域开始地址的相对位置：取值为 SEEK_SET 相对文件起始位置、SEEK_CUR 文件当前位置、SEEK_END 文件结束位置；

锁区域开始地址偏移量：同 l_whence 共同确定锁区域；

锁的长度：0 表示锁至文件末。

参数 cmd 的三种取值：F_GETLK、F_SETLK、F_SETLKW

例如：

将阻塞文件设置成非阻塞：

```
flags = fcntl(fd,F_GETFL,0);
flags |= O_NONBLOCK;
fcntl(fd,F_SETFL,flags);
```

fcntl 函数有五种功能：

复制一个现存的描述符，新文件描述符作为函数值返(cmd=F_DUPFD)。

获得/设置文件描述符标记（cmd = F_GETFD 或 F_SETFD），对应于 filedes 的文件描述符标志作为函数值返回。

获得/设置文件状态标志（cmd = F_GETFL 或 F_SETFL），对应于 filedes 的文件状态标志作为函数值返回。

获得/设置异步 I/O 有权（cmd = F_GETOWN 或 F_SETOWN）。

获得/设置记录锁（cmd = F_SETLK，F_SETLKW）。

例如：

```
void SeekLock(int fd,int start,int len)
{
  struct flock lock;
  lock.l_type=F_WRLCK;
  lock.l_whence=SEEK_SET;
  lock.l_start=start;
  lock.l_len=len;
  if(fcntl(fd,F_GETLK,&lock)==-1)
     fprintf(stderr,"See Lock failed./n");
  else if(lock.l_type==F_UNLCK)
     fprintf(stderr,"NO LOCK FROM %d TO %d,n",start, len);
  else if(lock.l_type==F_WRLCK)
     fprintf(stderr,"WRITE LOCK FROM %d TO %d,id=%d/n",start,len, lock.l_pid);
  else if(lock.l_type==F_RDLCK)
     fprintf(stderr,"READ LOCK FROM %d To %d,id=%d/n",start,len, lock.l_pid);
}
```

5.2 程序设计实例

实例 1 单词记忆小助手。设计一个程序，帮助单词记忆，此程序可以完成单词词义查找，单词随机输出。这些单词也可为用户自主定义为生词、难词、易错词。程序在主程序开始时应用 fork 函数创建子程序对单词数量进行计算，使程序同步在两个进程中运行，节省了运行的时间，通过信号的设置，在 reciew 函数中应用 ctrl+c 键进行退出，进入主界面，来结束无限循环。程序代码：

```c
#include<stdio.h>            //文件预处理，包含标准输入输出库
#include<unistd.h>           //文件预处理，包含fork,exit等函数
#include<stdlib.h>           //文件预处理，包含exit函数
#include<sys/types.h>        //文件预处理，包含pid_t定义
#include<sys/wait.h>         //文件预处理，包含wait等函数
#include<string.h>           //文件预处理，包含strcmp等函数
#include<signal.h>
void fun_ctrl_c();           //自定义信号处理函数声明
void add(FILE *fp,int *n)    //单词增加函数
{
 char s[80];
 if((fseek(fp,0,SEEK_END))==-1){   //使读写位置移动到文件尾
  perror("Add failed:");
  exit(1);
 }
 printf("Please enter the word:\n");  //输入单词
 fgets(s,80,stdin);
 fputs(s,fp);         //将单词存入文件
 fputc('\0',fp);
 printf("Please enter the meaning or notes:\n");
 fgets(s,80,stdin);           //输入词义，存入文件
 fputs(s,fp);
 fputc('\0',fp);
 (*n)=(*n)+2;
 printf("Totally %d words!\n",(*n)/2);   //输出现有单词个数
 fclose(fp);
 system("./dic");    //返回主界面
}
```

```c
void seek(FILE *fp,int *n)     //单词搜索函数
{
 int flag=0;
 char s[20],t[80],a;
 rewind(fp);          //使读写位置移到文件头
 printf("Please enter the word you want to seek:\n");
 fgets(s,20,stdin);      //输入要搜索单词
 while((fgets(t,80,fp))!=NULL){     //若不为文件尾
  if((strcmp(t,s))==0){
   flag=1;     //搜索到单词,退出循环
   break;
  }
 }
 if(flag==0){     //若flag为0则没有找到相应单词
  printf("Cannot find the word,would you like to add it?(y?)\n");
  scanf("%c",&a);
  getchar();
  if(a=='y') add(fp,n);     //若添加,进入添加函数
  fclose(fp);
  system("./dic");      //若不添加,返回主界面
 }
 else{          //若找到输出意思
  fgets(t,80,fp);
  printf("%s    :%s",s,t);
  fclose(fp);
  system("./dic");  //返回主界面
 }
}

void review(FILE *fp,int *n)     //复习函数
{
 (void)signal(SIGINT,fun_ctrl_c);   //如果按了ctrl+c键,调用fun_ctrl_c函数
 int i,m;
 char s[20];
 srand((int)time(0));      //设置随机种子
 while(1){
  i=(int)(1.0*(*n)*rand()/(RAND_MAX+1.0));     //随机输出比n小的数
  if((i%2)==0)  i++;       //保证其为奇数
```

```
  rewind(fp);
  for(m=0;m<i;m++){
   fgets(s,20,fp);
  }
  printf("%s",s);     //输出第i行数
  getchar();
 }
}

void fun_ctrl_c()   //退出函数
{
 char c;
 printf("Do you want to quit?\n");
 scanf("%c",&c);
 if(c=='y'){
  (void)signal(SIGINT,SIG_DFL);   //还原
  system("./dic");
  exit(0);
 }
 return;
}

int main(void)     //主函数
{
 char t;
 int cho,status,n,i=0;
 FILE *fp;
 pid_t result,wpid;
 if((fp=fopen("wlist","a+"))==NULL){   //打开文件
  perror("File open failed:");
  exit(1);
 }
 result=fork();    //建立子进程
 if(result<0){
  perror("Error:");
  exit(1);
 }
 else if(result==0){   //子进程
  rewind(fp);
```

```c
    while((t=fgetc(fp))!=EOF){
     if(t=='\n') i++;
    }
    printf("Totally %d words!\n",i/2);    //计算文件行数，获得单词数量
    exit(i);
  }
  else{
    printf("1-----Add new words\n2-----Seek a word\n3-----Review the words(random)\nPress any other key to quit\n");    //选择
    scanf("%d",&cho);
    getchar();
    wpid=wait(&status);    //等待子进程结束，防止出现僵尸进程
    n=WEXITSTATUS(status);    //获得子进程退出值，这里表示单词数量的两倍
    switch(cho){
     case 1:
      add(fp,&n);    //选择1 增加单词
      break;
     case 2:    //选择2 搜索单词
      seek(fp,&n);
      break;
     case 3:    //选择3 随机输出
      review(fp,&n);
      break;
     default:
      fclose(fp);
      printf("Thank you for using HYL dictionary!\n");
      sleep(1);
      break;
    }
  }
}
```

实例 2 设计一个程序。从键盘输入一个字符串作为文件名或路径名，程序判断输入的是否为目录，若为目录，则以字典顺序显示其下的文件及文件夹名，若非目录，则显示文件类型和大小。使用选择法排序的思想，让目录下的文件（夹）名以字典序显示。

源代码：

```c
#include <stdio.h>
#include <sys/stat.h>
#include <sys/types.h>
#include <unistd.h>
```

```c
#include <string.h>
#include <dirent.h>
#include <time.h>
/*目录处理函数*/
void f(char *p)
{
    DIR * dir;
    struct dirent * ptr;
    int n=0,i,j,index;
    char *temp,*file[512];
    /*打开目录*/
    dir=opendir(p);
    while((ptr = readdir(dir))!=NULL)
        /*将不以 . 开头的文件存入file数组*/
        if(*(ptr->d_name)!='.')
        {
            file[n]=ptr->d_name;
            n++;
        }
    closedir(dir);
    /*选择法排序*/
    for(i=0;i<n-1;i++)
    {
        index=i;
        for(j=i+1;j<n;j++)
            if(strcmp(file[j],file[index])<0)
                index=j;
        temp=file[i];
        file[i]=file[index];
        file[index]=temp;
    }
    for(i=0;i<n;i++)
        printf("%s\n",file[i]);
}
int main()
{
    struct stat buf;
    char path[512];
    printf("请输入一个路径:");
```

```c
    scanf("%s", path);
    /*用分支结构判断文件类型*/
    if(stat(path,&buf)<0)
        printf("该路径不存在\n");
    else if (S_ISLNK(buf.st_mode))
        printf("%s 是一个连接\n", path);
    else if (S_ISCHR(buf.st_mode))
        printf("%s 是一个字符设备\n", path);
    else if (S_ISDIR(buf.st_mode))
    {
        printf("%s 是一个目录\n", path);
        printf("%s 下的文件夹或文件有：\n", path);
            /*调用 f()输出文件名后即退出*/
        f(path);
        exit(0);
    }
    else if (S_ISREG(buf.st_mode))
        printf("%s 是一个常规文件\n", path);
    else if (S_ISBLK(buf.st_mode))
        printf("%s 是一个块设备\n", path);
    else if (S_ISFIFO(buf.st_mode))
        printf("%s 是一个 FIFO 文件\n", path);
    else if (S_ISSOCK(buf.st_mode))
        printf("%s 是一个 SOCKET 文件\n", path);
    else
        printf("未知类型文件\n");
    printf("文件大小：%d\n",buf.st_size);
}
```

实例 3 应用文件属性，编写一个程序，要求达到 Linux 中的命令 ls -l 功能，并且要求在格式上必须与系统的命令 ls 一致；即：-rw-r--r-- 1 root root 2915 08-03 06:16 a。

```c
#include <sys/types.h>
#include <sys/stat.h>
#include <unistd.h>
#include <stdio.h>
#include <stdlib.h>
#include <string.h>
#include<pwd.h>
#include<grp.h>
#include<time.h>
```

```c
void mode_to_letter(int mode,char *str)        //转换文件权限显示信息
{
    str[0]='-';        //缺省为'-'
    if(S_ISDIR(mode)) str[0]='d';              //判断 链接文件
    if(S_ISCHR(mode)) str[0]='c';              //判断 字符设备文件
    if(S_ISBLK(mode)) str[0]='b';              //判断 设备文件
    if(mode & S_IRUSR) str[1]='r';             //开始权限信息转换 用户
    else str[1]='-';
    if(mode & S_IWUSR) str[2]='w';
    else str[2]='-';
    if(mode & S_IXUSR) str[3]='x';
    else str[3]='-';
    if(mode & S_IRGRP) str[4]='r';             //开始权限信息转换 群组
    else str[4]='-';
    if(mode & S_IWGRP) str[5]='w';
    else str[5]='-';
    if(mode & S_IXGRP) str[6]='x';
    else str[6]='-';
    if(mode & S_IROTH) str[7]='r';             //开始权限信息转换 其他
    else str[7]='-';
    if(mode & S_IWOTH) str[8]='w';
    else str[8]='-';
    if(mode & S_IXOTH) str[9]='x';
    else str[9]='-';

    str[10]='\0';
}

int main(int argc,char *argv[])
{
    struct stat fst;                           //定义结构体 stat
    struct tm *mytime=(struct tm *)malloc(sizeof(struct tm));
    char str[12];
    if(argc!=2){
        fprintf(stderr,"Usage: %s <pathname>\n",argv[0]);
        exit(EXIT_FAILURE);
    }

    if(stat(argv[1],&fst)==-1){                //判断是否正确读取
```

```
        perror("stat");
        exit(EXIT_FAILURE);
    }


    mode_to_letter(fst.st_mode,str);        //开始文件权限模式信息修改
    printf("%s",str);                       //输出文件类型与权限信息
    printf(" %d",fst.st_nlink);             //输出文件的硬链接数
    printf(" %s",getpwuid(fst.st_uid)->pw_name);    //输出所属用户名
    printf(" %s",getgrgid(fst.st_gid)->gr_name);    //输出用户所在群组
    printf(" %ld",fst.st_size);             //输出文件大小
    mytime=localtime(&fst.st_mtime);        //获取文件修改时间
    printf("%d-%02d-%02d %02d : %02d", mytime -> tm_year + 1900, mytime -> tm_mon+1, mytime->tm_mday,mytime->tm_hour,mytime->tm_min);
        //按照一定格式输出文件修改时间
    printf("%s",argv[1]);                   //输出文件名
    printf("\n");                           //回车
    return 0;
}
```

第 6 章
进程控制

6.1 知识要点

6.1.1 进程调度命令 at

用户可使用 at 命令在指定时刻执行指定的命令序列。
语法格式：
at [-V] [-q 队列] [-f 文档名] [-mldbv] 时间

（1）使用 at 命令指定时间后按下回车，会出现 at>交互模式，在该模式中输入指令或程序，输入完毕后按下组合键 ctrl+D。执行结果将会被寄回账号。

（2）可以先将所有的指令先写入档案后，利用-f 一次性读入。

例如：

现在时间是 2012 年 8 月 1 日 中午 12:00，要求在今天下午 3:00 执行某命令，可用以下方式来表示时间：

1）at 3:00pm
2）at 15:00
3）at now + 3 hours
4）at 15:00 8/1/12

例如：

找出系统中以.c 为后缀名的文件并打印，结束后给用户 yin 发出一封名叫 done 的邮件，指定时间为 8 月 1 日中午 12 点。

```
$ at 12:00 8/1/12
```

回车后系统出现 at>提示符，输入命令：

```
at> find / -name "*.c" | lpr
at> echo "yin: work is finished!" | mail -s "done" yin
```

回车后按组合键 ctrl+D 结束输入。系统会提醒用户将使用哪个 shell 来执行该命令序列，如：

```
warning:command will be executed using /bin/sh
job 1 at 2012-08-01 12:00
```

指定时间技巧：

1) 可接受 hh:mm 的指定，如果该时间已经过去，就放在第二天执行；
2) 可接受 12 小时计时，在时间后加上 am 或 pm 来指明上午或下午；
3) 可接受具体日期，如 month day 或 mm/dd/yy 或 dd.mm.yy；
4) 可接受口语词汇如 midnight\noon\teatime 等；
5) 可接受相对计时法，格式为 now + count time units，即当前时间加上时间与单位，单位可为 minutes\hours\days\weeks 等；
6) 可接受 today\tomorrow。

6.1.2 进程创建函数 fork

调用 fork()函数一次分别在父进程、子进程各返回一个值，在子进程中返回值为 0，在子进程中返回的是父进程的进程号。通常用 fork 函数的程序设计格式为：

```
childpid= fork();
if(childpid < 0)
{
        //子进程产生失败
}
else if(childpid == 0)
{
        //子进程
}
else
{
        //父进程
}
```

例如：
```
childpid= fork();
 if(childpid < 0)
{
    printf("fail\n");
}
else if(childpid == 0)
{
    printf("son\n");
}
else
{
    printf("father");
    retpid = waitpid(childpid,&status,0);
    if(retpid == childpid)
        printf("son finished, ready to start father...\n");
}
```

6.1.3 僵尸进程

在一个子进程已经结束，但它的父进程没有对它进行及时的善后处理（获取终止子进程的有关释放信息但仍占用的资源）的进程被称为僵尸进程(zombie)。僵尸进程虽然已经结束，但是在进程表项中仍然占用系统可用的程序号。为了免除这种危害，需要使用ps、top和kill命令来查杀僵尸进程，利用wait和waitpid函数暂停父进程也可以预防僵尸进程的产生。同时要注意，在使用kill命令停止父进程时，要先将子进程逐一停止。

进程在它的生命周期有睡眠、运行、停止和僵死等状态。一个进程退出时，它的父进程会收到一个SIGCHLD信号。一般情况下，这个信号的句柄通常执行wait系统调用，这样处于僵死状态的进程会被删除。如果父进程没有这么做，进程会处于僵死状态。

那么如何处理掉僵尸进程呢？补救办法是杀死僵尸进程的父进程(僵尸进程的父进程必然存在)，这时僵尸进程会成为"孤儿进程"，过继给1号进程init，init始终会负责清理僵尸进程。

例如：
```
pc=fork( );
if(pc==0)
printf("child");
else
{
sleep(20);
```

```
        printf("parent");
    }
```

则会由于父进程的休眠导致僵尸进程的产生。

使用wait函数可以作用于父进程,使得其在其子进程结束或者收到特殊信号前不结束进程,从而可以有效防止僵尸进程的产生。

6.1.4 wait 与 waitpid 函数

1.函数.pid_t   wait（int status）

进程一旦调用了 wait,就会立即阻塞自己,由 wait 自动分析是否当前进程的某个子进程已经退出。如果让它找到了一个已经变成僵尸的子进程,wait 就会收集这个子进程的信息,并把它彻底销毁后返回;如果没有找到这样一个子进程,wait 就会一直阻塞在这里,直到有一个已经变成僵尸的子进程出现为止。

如果对这个子进程是如何被杀死的毫不在意,只想把这个僵尸进程消灭掉,可以设定这个参数为 NULL,如:

pid = wait(NULL);

如果成功,wait 会返回被收集的子进程的进程 ID,如果调用进程没有子进程,调用就会失败,此时 wait 返回-1,同时 errno 被置为 ECHILD。

2.函数 waitpid(pid,status,options)

当 pid 取不同的值时,有不同的意义:

pid>0 时,只等待进程 ID 等于 pid 的子进程,不管其它已经有多少子进程运行结束退出,只要指定的子进程还没有结束,waitpid 就会一直等下去。

pid=-1 时,等待任何一个子进程退出,没有任何限制,此时 waitpid 和 wait 的作用一模一样。

pid=0 时,等待同一个进程组中的任何子进程,如果子进程已经加入了别的进程组,waitpid 不会对它做任何理睬。

pid<-1 时,等待一个指定进程组中的任何子进程,这个进程组的 ID 等于 pid 的绝对值。

options 提供了一些额外的选项来控制 waitpid,目前在 Linux 中只支持 WNOHANG 和 WUNTRACED 两个选项,这是两个常数,可以用"|"运算符把它们连接起来使用。比如:

ret=waitpid(-1,NULL,WNOHANG | WUNTRACED);

如果我们不想使用它们,也可以把 options 设为 0,如:

ret=waitpid(-1,NULL,0);

6.1.5 僵尸进程的避免

程序中程序语句 wpid=wait（&status）；起着使父进程调用 wait 函数，消除僵尸进程的作用。在此处，用程序语句 wpid=waitpid（pid，&status，0）；也能发挥同等作用。借助这两个函数，就可以防止子进程在父进程早于子进程终止时成为僵尸进程。

为了避免僵尸进程的产生，父进程要等待子进程，而子进程结束时通知父进程，最后父进程退出，程序结构可以设计为：

```
do{
    pr=waitpid(pc, NULL, WNOHANG);   /* 因为 WNOHANG, waitpid 不予等待 */
    if(pr==0)                        /* 如果没有收集到子进程 */
     {
      printf("父进程未捕捉到子进程终止\n");
      sleep(1);                      /*父进程等待1 秒 */
     }
}while(pr==0);                       /*父进程未捕捉到子进程终止,执行循环*/
if(pr==pc)
{
  printf("父进程捕捉到子进程终止,子进程的进程号(PID)是:%d\n",pr);
}
else
  printf("父进程无法捕捉到子进程终止\n");
```

6.1.6 守护进程

守护进程编写要点：
1.结束父进程，子进程继续
代码：
```
if( (pid = fork()) > 0)
    exit(0);       //结束父进程,
```
2.脱离控制终端和进程组
代码：
```
setsid();   //脱离控制终端
```
setsid 有三个作用：让进程摆脱原会话、原进程组、原控制终端的控制。
3.改变当前工作目录
由于在进程运行过程中，当前目录所在的文件系统不能卸载。因此，要把当前工作目录换成其他的路径。代码：

```
chdir("/");
```

4.重设文件创建掩码

进程从创建它的父进程那里继承了文件创建掩模。它可能修改守护进程所创建的文件的存取位。为防止这一点，将文件创建掩模清除。代码：

```
umask(0);
```

5.关闭打开的文件描述符

进程从创建它的父进程那里继承了打开的文件描述符。如不关闭，将会浪费系统资源，造成进程所在的文件系统无法卸下以及引起无法预料的错误。代码：

```
#define MAX_FILENO 1024
for(i=0;i<MAX_FILENO;++i)
   close(i);
```

例如：

产生守护进程的程序代码段。

```
    child=fork();           // 创建子进程，终止父进程，子进程在后台执行
    if(child>0)
        exit(0);
    else if(child<0)
    {
        perror("创建子进程失败！");
        exit(1);
    }
    setsid();       // 在子进程中创建会话
    chdir("/tmp");                         // 改变工作目录到"/tmp"
    umask(0);       // 重设文件创建掩码
    for(i=0;i< NOFILE;++i)                 //关闭文件描述符
    close(i);
```

6.2 程序设计实例

实例 1 程序的目的是避免产生僵尸进程，避免产生僵尸进程的方法有很多，例如课本上的方法是用 fork 函数复制进程，返回值为 0 的为子进程，大于 0 的为父进程，父进程用 wait 函数等待子进程正常终止。在本程序中则选择使用由父进程调用 signa 函数将 SIGCHLD 的处理动作置为 SIG_IGN，这样 fork 出来的子进程在终止时也会自动清理掉的方法。程序中用到了 waitpid, sleep, signal, fork 等函数。

```c
#include <stdio.h>
#include <unistd.h>
#include <stdlib.h>
#include <signal.h>
#include <sys/types.h>
#include <sys/wait.h>
int main()
{
pid_t p;
        int i = 0, j = 0;
        p = fork();
        if (p<0) {
                perror("fork error\n");
                exit(1);
        }
        if (p>0) {
            for (; i<20; i++) {
                sleep(1);
                signal(SIGCHLD, sig_term);
                printf("parent\n");
            }
        }
else{
            for (; j<5; j++) {
                sleep(1);
                printf("child\n");
            }
            exit(2);
        }
        return 1;
}

void sig_term(int sig)
{
   pid_t p;
   int i = 0;
   waitpid(p, &i, 0);
   printf("%d\n",WEXITSTATUS(i));
}
```

实例 2　创建一个可以每隔一段时间改变本机 IP 地址的守护进程,并把所更改的 IP 地址记录在了 IPLOG 文件中。程序中涉及流文件的操作、系统函数的使用、守护进程的创建等,并具有一定的实用性。程序代码为:

```
#include<stdio.h>
#include<time.h>
#include<unistd.h>
#include<signal.h>
#include<sys/param.h>
#include<sys/types.h>
#include<sys/stat.h>
#include<stdlib.h>
#include<sys/ioctl.h>
#include<string.h>
void init_daemon(void);
int main()
{
char IP[13], cmd[100] = "ifconfig eth0 ";
FILE *fp,*fp1;
int offset;
init_daemon();
while(1)
{
    if((fp = fopen("IP","r")) == NULL)
    {
        printf("failed to open IP");
        exit(1);
    }
    if((fp1 = fopen("IPLOG","a+")) == NULL)
    {
        printf("failed to open IPLOG");
        exit(1);
    }
    srand((int)time(0));
    offset = rand()/(RAND_MAX+1.0)*10.0;
    fseek(fp, offset*13, SEEK_SET);
    fread(IP, 13, 1, fp);
    strcat(cmd,IP);
    fprintf(fp1," 正在运行%s",cmd);
```

```c
        system(cmd);
        //execlp("ifconfig","ifconfig","eth0",IP,NULL);
        rewind(fp);
        strcpy(cmd,"ifconfig eth0 ");
        fclose(fp);
        fclose(fp1);
        sleep(30);
    }
}
//创建守护进程
void init_daemon(void)
{
    pid_t child1;
    int i;
    child1 = fork();
    if(child1 > 0)
        exit(0);
    else if(child1 < 0)
    {
        perror("failed");
        exit(1);
    }
    setsid();                          // 创建新会话期
    chdir("/home/nearu/c_prog");       // 改变更目录
    umask(0);                          // 重置文件掩码
    for(i = 0; i < NOFILE; i++)        // 关闭文件描述符
        close(i);
    return;
}
```

实例 3 编写守护进程，检查邮箱并提醒的程序。要求每隔五分钟检查一次邮箱，并记录在一个 **/tmp** 的 mail.log 文件里，有新邮件时警报提醒。mail.log 文件记录有检查时间、检查时存在的文件数及列表等。提示可以在演示时把检查邮箱的间隔时间改为 10 秒，以在 /var/spool/mail 内新建文件来模仿收到新邮件。程序代码为：

```c
#include<dirent.h>
#include<time.h>
#include<stdio.h>
#include<stdlib.h>
#include<sys/types.h>
#include<unistd.h>
```

```c
#include<sys/wait.h>
#include<syslog.h>
#include <signal.h>
#include <sys/param.h>
#include <sys/stat.h>
int main()
{
  pid_t child1;
  FILE *fp;
  int i;
  int flag,nf,nnf;
  int nfile();
  time_t t;
  child1=fork();
  if(child1>0)
    exit(0);
  else if(child1< 0)
  {
    perror("创建子进程失败");
    exit(1);
  }
  setsid();
  chdir("/tmp");
  umask(0);
  for(i=0;i<NOFILE;++i)
    close(i);
nf=nfile(fp);
while(1)
{
  sleep(300);
  fp=fopen("mail.log","a+");
  if(fp<0)
      perror("文件打开失败。");
  nnf=nfile(fp);
  if(nnf>nf){          //通过判断文件数量变化来判定新邮件
      fprintf(fp,"您有新邮件请注意查收！\n");
      putchar(7);
  }
  else{
```

```
        fprintf(fp,"此时间段没有新邮件。\n",nf,nnf);
    }
    t=time(0);
    fprintf(fp,"%s",asctime(localtime(&t)));
    fclose(fp);
    nf=nnf;
  }
}
int nfile(FILE *fp)  //计算文件夹内文件数量
{
    DIR *dir;
    struct dirent *ptr;
    int i=0;
    dir=opendir("/var/spool/mail");
    while((ptr=readdir(dir))!=NULL)
    {
        i++;
        fprintf(fp,"%s\n",ptr->d_name);
    }
    closedir(dir);
    return i;
}
```

实例 4 文件监视记录程序,此程序通过创建一个守护进程,来监视一个选定的文件或文件夹。如果文件或文件夹发生了改动,如创建了新的文件或删除了文件,则会把改动文件的详细情况记录在 **/tmp/history/** 下的 **add.txt** 和 **deleted.txt** 里。该程序可以应用在监视一些重要的文件夹,如系统文件夹之类的,或者重要信息资料的文件夹。

1. Inspector.c 程序代码

```
#include <stdio.h>
#include <time.h>
#include <unistd.h>
#include <sys/stat.h>
#include <string.h>
void create_daemon();
void compare_lost();
void compare_add();
int main()
{
    struct stat buf;
    char filename[1000],command[1000],temp[]="ls ";
```

```c
    int i;
    int static flag;
    printf("请输入想要监管的文件或文件夹：\n");
    scanf("%s",filename);
    stat(filename,&buf);
    if (S_ISDIR(buf.st_mode))
    {
        printf("输入文件夹地址有效\n");
    }
    else if (S_ISREG(buf.st_mode))
    {
        printf("输入文件地址有效\n");
    }
    else
    {
        perror("无效地址！\n");
        exit(0);
    }
    create_daemon();
    strcat(temp,filename);
    strcpy(command,temp);
    strcat(temp," -l >> /tmp/history/last.txt");
    system("mkdir /tmp/history");
    system(temp);
    while (1)
    {
        system("rm /tmp/history/latest.txt");
        strcat(command," -l >> /tmp/history/latest.txt");
        system(command);
        compare_lost();
        compare_add();
        sleep(10);
    }
    scanf("%d",&i);
}
```

2.compare.c 程序代码

```c
#include <stdio.h>
#include <stdlib.h>
```

```c
#include <string.h>
#include <unistd.h>
#include <sys/stat.h>
void compare_lost()
{
    int size1,size2,flag3=1,flag4=1;
    int static flag1;
    char tmp1[1000],tmp2[1000],tmp3[1000];
    struct stat buf1,buf2;
    FILE *fp1, *fp2, *fp3;
    stat("/tmp/history/last.txt",&buf1);
    stat("/tmp/history/latest.txt",&buf2);
    size1=buf1.st_size;
    size2=buf2.st_size;
    if (size1!=size2)
    {
        if ((fp1=fopen("/tmp/history/last.txt","r"))==NULL ||
            (fp2=fopen("/tmp/history/latest.txt","r"))==NULL ||
            (fp3=fopen("/tmp/history/deleted.txt","a+"))==NULL)
        {
            exit(0);
        }
        while (!feof(fp1))
        {
            fgets(tmp1,100,fp1);
            flag3=1;
            while (!feof(fp2))
            {
                fgets(tmp2,100,fp2);
                if (strcmp(tmp1,tmp2)==0)
                {
                    flag3=0;
                    break;
                }
            }
            if (flag3)
            {
                fclose(fp3);
                if ((fp3=fopen("/tmp/history/deleted.txt","r"))==NULL)
```

```c
                exit(0);
                  flag4=1;
                  while (!feof(fp3))
                  {
                      fgets(tmp3,100,fp3);
                      if (strcmp(tmp1,tmp3)==0)
                      {
                          flag4=0;
                          break;
                      }
                  }
                  fclose(fp3);
                  if((fp3=fopen("/tmp/history/deleted.txt","a+"))==NULL)
                      exit(0);
                  if (!flag1)
                  {
                      fputs("丢失的文件：\n",fp3);
                      flag1=1;
                  }
                  if (flag4 && (tmp1[0]!='t' && tmp1[1]!='o'))
                  {
                      fputs(tmp1,fp3);
                  }
              }
              fseek(fp2,0,0);
          }
          fclose(fp1);
          fclose(fp2);
          fclose(fp3);
      }
}
void compare_add()
{
    int size1,size2,flag3=1,flag4=1;
    int static flag2;
    char tmp1[1000],tmp2[1000],tmp3[1000];
    struct stat buf1,buf2;
    FILE *fp1, *fp2, *fp3;
    stat("/tmp/history/last.txt",&buf1);
```

```c
        stat("/tmp/history/latest.txt",&buf2);
        size1=buf1.st_size;
        size2=buf2.st_size;
        if (size1!=size2)
        {
            if ((fp1=fopen("/tmp/history/latest.txt","r"))==NULL ||
                (fp2=fopen("/tmp/history/last.txt","r"))==NULL ||
                (fp3=fopen("/tmp/history/new.txt","a+"))==NULL)
            {
                exit(0);
            }
            while (!feof(fp1))
            {
                fgets(tmp1,100,fp1);
                flag3=1;
                while (!feof(fp2))
                {
                    fgets(tmp2,100,fp2);
                    if (strcmp(tmp1,tmp2)==0)
                    {
                        flag3=0;
                        break;
                    }
                }
                if (flag3)
                {
                    fclose(fp3);
                    if ((fp3=fopen("/tmp/history/new.txt","r"))==NULL) exit(0);
                    flag4=1;
                    while (!feof(fp3))
                    {
                        fgets(tmp3,100,fp3);
                        if (strcmp(tmp1,tmp3)==0)
                        {
                            flag4=0;
                            break;
                        }
                    }
                    fclose(fp3);
```

```
                if        ((fp3=fopen("/tmp/history/new.txt","a+"))==NULL)
exit(0);
                if (!flag2)
                {
                    fputs("新增的文件：\n",fp3);
                    flag2=1;
                }
                if (flag4 && (tmp1[0]!='t' && tmp1[1]!='o'))
                {
                    fputs(tmp1,fp3);
                }
            }
            fseek(fp2,0,0);
        }
        fclose(fp1);
        fclose(fp2);
        fclose(fp3);
    }
}
```

3.create.c 程序代码

```
#include <unistd.h>
#include <signal.h>
#include <sys/param.h>
include <sys/types.h>
#include <sys/stat.h>
void create_daemon()
{
    pid_t p1,p2;
    int i;
    p1=fork();
    if (p1<0)
    {
        perror("创建子进程失败！\n");
        exit(1);
    }
    else if (p1>0)
    {
        printf("父进程结束！\n");
        printf("守护进程开启！\n");
```

```c
        exit(0);
    }
    setsid();
    chdir("/tmp");
    umask(0000);
    for (i=0;i<NOFILE;i++)
        close(i);
    return;
}
```

第 7 章 进程间的通信

7.1 知识要点

7.1.1 Linux 进程间的通信方式

Linux 进程间的通信方式有：①无名管道（pipe）和有名管道（FIFO）；②信号（signal）；③消息队列；④共享内存；⑤信号量；⑥套接字（socket）。

7.1.2 进程间通信的特点

（1）无名管道：数据只能在父子进程间单向流动。管道是半双工的，数据只能向一个方向流动；需要双方通信时，需要建立起两个管道；只能用于父子进程或者兄弟进程之间（具有亲缘关系的进程）

（2）有名管道：允许非父子进程的数据流动。由于命名管道相比无名管道，其不同之处在于添加了一个名字，因此命名管道相当于在文件系统之下建立了一个可以被访问的文件，所以命名管道相比无名管道在操作上也多了一个步骤，即打开（open）管道操作。

（3）信号：用于通知进程事件的发生，常用 signal 函数来设置信号的处理方式、返回先前的信号处理函数指针。

(4) 消息队列:实质为消息链表存放于内核并用标识符标识,与管道、信号相比,能传输种类更多、信息量较大的信号。发送方不必等待接收方检查它所收到的消息就可以继续工作下去,而接收方如果没有收到消息也不需等待。

(5) 共享内存:通过映射内存的方式,使得一段内存能被创建进程在内的多个进程访问,直接而效率高,通常和其他方式配合使用。

(6) 信号量:用于控制多进程对共享资源的访问,多用于进程的同步。

(7) 套接字(socket):套接字也是一种进程间通信机制,与其他通信机制不同的是,它可用于网络中不同机器之间的进程间通信。

7.1.3 管道通信的函数

1.管道函数 int pipe(int fd[2])

该函数创建的管道的两端处于一个进程中间,一个进程在由 pipe()创建管道后,一般再 fork 一个子进程,然后通过管道实现父子进程间的通信。管道两端可分别用描述字 fd[0]以及 fd[1]来描述,需要注意的是,管道的两端是固定了任务的。即一端只能用于读,由描述字 fd[0]表示,称其为管道读端;另一端则只能用于写,由描述字 fd[1]来表示,称其为管道写端。一般文件的 I/O 函数都可以用于管道,如 close、read、write 等。

注意:

(1) 管道是一个固定大小的缓冲区,在 Linux 中该缓冲区的大小为1页,即 4K 字节,使得它的大小不象文件那样不加检验地增长。使用单个固定缓冲区也会带来问题,比如在写管道时可能变满,当这种情况发生时,随后对管道的 write()调用将默认地被阻塞,等待某些数据被读取,以便腾出足够的空间供 write()调用写。

(2) 读取进程也可能工作得比写进程快。当所有当前进程数据已被读取时,管道变空。当这种情况发生时,一个随后的 read()调用将默认地被阻塞,等待某些数据被写入,这解决了 read()调用返回文件结束的问题。

2.管道的读写规则

管道函数 pipe 创建的管道的两端处于一个进程中间,在实际应用中没有太大意义,因此,一个进程在由 pipe()创建管道后,一般再 fork 一个子进程,然后通过管道实现父子进程间的通信。管道两端可分别用描述字 fd[0]以及 fd[1]来描述,管道读端:fd[0],管道写端:fd[1]。

(1) 读管道规则

关闭管道的写端:close (fd[WRITE]);

读出:read(fd[READ], string, strlen(string));

读出后关闭管道的读端:close(fd[READ]);

(2) 写管道规则

关闭管道的读端:close(fd[READ]);

写入：write(fd[WRITE], string, strlen(string));
写入后关闭管道的写端：close (fd[WRITE]);

7.1.4 信号传送和处理

1.信号源

进程通信主要用于用户进程和系统内核进程间通信，信号是进程间通信的方法之一，信号可以在任何时候发给某一进程。信号发生有两个来源：

（1）硬件来源：按下了 Delete 或用鼠标单击，通常产生中断信号；

（2）软件来源：通过系统调用或使用命令发出信号，或一些非法的操作。

常见信号 SUGHUP、SIGINT、SIGQUIT、SIGILL、SIGFPE、SIGKILL、SIGALRM、SIGSTOP、SIGTSTP、SIGCHLD

2.信号阻塞程序的设计

首先初始化一个信号集，然后把信号加入信号集，最后把信号集合加入到当前进程的阻塞集合，程序段可以设计步骤为：

（1）初始化信号集合 sigemptyset(&set);

（2）设置信号处理方式 signal(signum,function);

（3）把信号加入信号集合 sigaddset(&set,signum);

（4）把信号集合加入到当前进程的阻塞集合中 sigprocmask(SIG_BLOCK,&set,NULL)。

例如：

```
if(sigemptyset(&set)<0)                  /*初始化信号集合*/
    printf("初始化信号集合错误");
if(sigaddset(&set,SIGINT)<0)              /*把SIGINT信号加入信号集合*/
    printf("加入信号集合错误");
if(sigprocmask(SIG_BLOCK,&set,NULL)<0)/*把信号集合加入到当前进程的阻塞集合中
*/
    printf("往信号阻塞集增加一个信号集合错误");
else
{
//主程序完成的工作
}
```

如果要处理此信号，需要对此信号解除阻塞，程序语句写为：

```
if(sigprocmask(SIG_UNBLOCK,&set,NULL)<0)  /*当前的阻塞集中删除一个信号集合*/
    printf("从信号阻塞集删除一个信号集合错误");
```

例如：

自动关机的小程序。运行程序后即自动关机。

```
#include <signal.h>
#include <stdio.h>
```

```c
#include <unistd.h>
#include <sys/reboot.h>
int main(int argc, char **argv)
{
    /* 先关闭所有信号 */
    sigset_t set;
    sigfillset(&set);
    sigprocmask(SIG_BLOCK, &set, NULL);
    /* 发送信号给所有进程*/
    printf("发送终止进程信号给所有进程\n");
    kill(-1, SIGTERM);
    sync();
    sleep(3);
    printf("发送终止进程信号给所有进程\n");
    kill(-1, SIGKILL);
    sync();
    sleep(3);
    /* 关机 */
    printf("系统关机\n");
    sleep(2);
    reboot(RB_POWER_OFF);
}
```

7.1.5 消息队列应用

消息队列是 Unix 系统 V 版本中 3 种进程间通信机制之一。消息队列就是一个消息的链表。把消息看作一个记录，并且这个记录具有特定的格式以及特定的优先级。对消息队列有写权限的进程可以按照一定的规则添加新消息；对消息队列有读权限的进程则可以从消息队列中读出消息。Linux 采用消息队列的方式来实现消息传递。消息的发送方式是：发送方不必等待接收方检查它所收到的消息就可以继续工作下去，而接收方如果没有收到消息也不需等待。新的消息总是放在队列的末尾，接收的时候并不总是从头来接收，可以从中间来接收。消息队列是随内核持续的并和进程无关，只有在内核重起或者显式删除一个消息队列时，该消息队列才会真正被删除。因此系统中记录消息队列的数据结构位于内核中，系统中的所有消息队列都可以在结构 msg_ids 中找到访问入口。

1.调用 ftok 接口产生一个 key

使用 ftok 的好处是，访问同一个消息队列的不同进程可以通过同一个文件访问相同的队列。另外，如果文件被删除，即使重新产生内容一模一样的新文件，仍然产生不同的 key，因为文件的 inode 与 key 值的产生有关。

2.产生一个消息队列

应用 key 作为 msgget 函数参数,产生一个消息队列。

3.发送与接收信息

进程可以用 msgsnd 发送消息到这个队列,相应的别的进程用 msgrcv 读取。注意 msgsnd 可能会失败的两个情况:A) 可能被中断打断(包括 msgsnd 和 msgrcv),B) 消息队列满。

4.msgctl 函数可以用来删除消息队列

消息队列产生之后,除非明确的删除,产生的队列会一直保留在系统中。Linux 下消息队列的个数是有限的,注意不要泄露。如果使用已经达到上限,msgget 调用会失败,产生的错误码对应的提示信息为 no space left on device。

例如:

```c
if((key==ftok(KEYPATH,PROJECT_ID))==-1)
{
 perror("Error in creating the key:");
 exit(1);
 }
if((msgpid=msgget(key,IPC_CREAT | 0666))==-1)
{
    perror("Error in creaing the message guene:");
    exit(1);
 }
 ...........
 pid1=fork();
if(pid1<0){
   perror("Error:");
   exit(1);
}
else if(pid1==0){
    while(1){
     len=msgrcv(msgpid,&msgr,sizeof(msgr),msgs.mtype,0);
     if(len==0)
       continue;
     if((strcmp(msgr.name,msgs.name))==0){
       msgsnd(msgpid,&msgr,100,0);
       continue;
     }
    }
}
```

7.1.6 共享内存函数 mmap 应用

mmap 系统调用使得进程之间通过映射同一个普通文件实现共享内存,mmap 把磁盘文件的一部分直接映射到内存,进程可以像读写内存一样对普通文件进行操作。如果 mmap 成功则返回映射首地址,如果出错则返回常数 MAP_FAILED。

```
void *mmap(void *addr,      //通常取 NULL 内核会自己在进程地址空间中建立映射
    size_t len,      //需要映射的那一部分文件的长度
    int prot,    //
    int flag,    //
    int filedes,    //文件的描述符
    off_t off);    //从文件的什么位置开始映射
```

prot 参数有四种取值:
PROT_EXEC 表示映射的这一段可执行,例如映射共享库
PROT_READ 表示映射的这一段可读
PROT_WRITE 表示映射的这一段可写
PROT_NONE 表示映射的这一段不可访问

flag 参数取值:
MAP_SHARED 多个进程对同一个文件的映射是共享的,一个进程对映射的内存做了修改,另一个进程也会看到这种变化。
MAP_PRIVATE 多个进程对同一个文件的映射不是共享的,一个进程对映射的内存做了修改,另一个进程并不会看到这种变化,也不会真的写到文件中去。

例如:
```c
#include <stdlib.h>    // 设程序名为 kk.c
#include <sys/mman.h>
#include <fcntl.h>
int main(void)
{
 int *p;
 int fd = open("Liu", O_RDWR);
 if (fd < 0) {
  perror("open hello");
  exit(1);
 }
 p = mmap(NULL, 6, PROT_WRITE, MAP_SHARED, fd, 0);
 if (p == MAP_FAILED) {
  perror("mmap");
  exit(1);
```

```
    }
    close(fd);
    p[0] = 0x30313233;
    munmap(p, 6);
    return 0;
}
```

1)建立文件 liu,并设内容为 Linux

2)应用命令 od 查看 liu 文件的内容,显示如下:

```
[root@localhost root]# od -tx1 -tc liu
0000000  4c  69  6e  75  78  0a
          L   i   n   u   x  \n
```

3)执行程序

```
[root@localhost root]# ./kk1
```

4)再次应用命令 od 查看 liu 文件的内容,显示如下:

```
[root@localhost root]# od -tx1 -tc liu
0000000  33  32  31  30  78  0a
          3   2   1   0   x  \n
0000006
```

5)你也可以打开文件 liu,发现 liu 的内容确已改写为 3210x

7.2 程序设计实例

实例1 创建一个管道 pipe,应用函数 fork 复制子进程,在父进程中运行命令"ls -1",把运行结果写入管道,子进程从管道中读取"ls -1"的结果,把读出的作为输入。程序代码:

```
#include<stdio.h>
#include<stdlib.h>
#include<sys/types.h>
#include<sys/wait.h>
#include<unistd.h>

int main ()
{
pid_t result;
int num;
```

```c
    int pipe_fd[2];
    FILE *fp;
    char buf[500];
    memset(buf,0,sizeof(buf));
    printf("Creating tunnel... ...\n");
    if(pipe(pipe_fd)<0) {
        printf("Can't create a tunnel");
        return -1;
    }
    result=fork();
    if(result<0) {
        perror("Can't create a process");
        exit;
    }
    else if (result==0) {
        close(pipe_fd[1]);
        if((num=read(pipe_fd[0],buf,500))>0)
        printf("The result from son process: %s\n",buf);
        close(pipe_fd[0]);
        exit(0);
    }else{
        close(pipe_fd[0]);
        fp=popen("ls -l","r");
        num=fread(pipe_fd[1],sizeof(char),500,fp);
        pclose(fp);
        close(pipe_fd[1]);
        waitpid(result,NULL,0);
        exit(0);
    }
    return 0;
}
```

实例 2 该程序通过用本地文件存储留言信息实现留言内容的保存，使得对象用户在下次打开时就能查看他人的留言信息。此留言板的功能有：用户可以对指定用户名进行留言、用户在输入用户名后查看自己的留言箱、用户可以选择清空留言箱内容、留言箱支持显示用户名、留言时间，本程序还应用了信号 **Ctrl+C** 中断来执行退出函数。程序中留言板的内容读取与存储，实质是对文件的读写操作；用 **strftime** 函数来简略读取时间；用 **remove** 函数来清空邮箱（实质是删除留言文件）；当用户按下 **Ctrl+C** 时执行中断，进入退出函数，并提示清空邮箱消息。程序代码为：

```c
#include<stdio.h>
```

```c
#include<string.h>
#include<stdlib.h>
#include<sys/types.h>
#include<sys/stat.h>
#include<errno.h>
#include<unistd.h>
#include<sys/time.h>
#include<time.h>
#include<signal.h>

FILE *fp1,*fp2;
char name1[100],name2[100];                    /*留言用户名*/
char s[300];              /*留言字符串*/
/******************************************************************
函数功能：获取当前时间，并返回hh:mm:ss格式的时间字符串
返回值：一个字符指针，指向时间信息
******************************************************************/
char* gettime(char time[8])
{
struct timeval tms;
char tstr[8];
timerclear(&tms);
gettimeofday(&tms, NULL);
strftime(time,100,"%X",localtime(&tms.tv_sec));
return time;
}


/******************************************************************
          函数功能：清空留言板
******************************************************************/
quitmsg()
{
    char choise;
    printf("是否清空您的留言箱(Y/N):");
    scanf("%c",&choise);
    if(choise=='Y')              /*用户选择清空,则清空留言板*/
    {
     remove(name1);
    }
```

```c
        printf("\n 感谢使用！\n");
        fclose(fp1);
        exit(0);
}

int main()
{
        signal(SIGINT,quitmsg);
        char s1[300],s2[200];
        char time[8];
        int flag=0;                    /*用于标记留言箱是否有留言*/
        printf("请输入你的名字：");
        scanf("%s",name1);
        if((fp2=fopen(name1,"a+"))==NULL)
        {
            printf("打开（创建）文件出错");
            exit(0);
        }
        printf("以下是您留言箱中的内容：\n");
        while(!feof(fp2))
        {
         if(fgets(s1,300,fp2)!=NULL)
         {
           puts(s1);
         }
        }
fclose(fp2);
printf("\n 请输入您要对其留言的用户名:");        /*输入对象用户名*/
scanf("%s",name2);
printf("\n 请输入您的留言内容，按 Ctrl+C 退出:\n");
if((fp1=fopen(name2,"a+"))==NULL)               /*打开留言对象留言文件*/
{
  printf("打开（创建）文件出错");
  exit(0);
}
while(1)
{
  strcpy(s,name1);
  strcat(s,":");
```

```
        scanf("%s",s2);
        gettime(time);
        strcat(s," ");
        strcat(s,s2);
        strcat(s,"   ");
        strcat(s,time);                    /*留言末加上时间字符串*/
        strcat(s,"\n");
        if(fputs(s,fp1)==EOF)
        {
         printf("写入文件错误！");
         exit(1);
        }
   }
 }
```

实例3 程序的功能是探索消息队列的容量，对消息队列属性作一个较为全面的了解。

源代码：

```
    #include <sys/types.h>
    #include <sys/msg.h>
    #include <unistd.h>
    void msg_stat(int,struct msqid_ds );
    main()
    {
      int gflags,sflags,rflags;
      key_t key;
      int msgid;
      int reval;
      struct msgsbuf
      {
       int mtype;
       char mtext[1];
       } msg_sbuf;
      struct msgmbuf
      {
          int mtype;
          char mtext[10];
      }msg_rbuf;
      struct msqid_ds msg_ginfo,msg_sinfo;
      char* msgpath="/unix/msgqueue";
      key=ftok(msgpath,'a');
```

```
gflags=IPC_CREAT|IPC_EXCL;
msgid=msgget(key,gflags|00666);
if(msgid==-1)
{
    printf("msg create error\n");
    return;
}
msg_stat(msgid,msg_ginfo);
sflags=IPC_NOWAIT;
msg_sbuf.mtype=10;
msg_sbuf.mtext[0]='a';
reval=msgsnd(msgid,&msg_sbuf,sizeof(msg_sbuf.mtext),sflags);
if(reval==-1)
{
    printf("message send error\n");
}
msg_stat(msgid,msg_ginfo);
rflags=IPC_NOWAIT|MSG_NOERROR;
reval=msgrcv(msgid,&msg_rbuf,4,10,rflags);
if(reval==-1)
    printf("read msg error\n");
else
    printf("read from msg queue %d bytes\n",reval);
msg_stat(msgid,msg_ginfo);
msg_sinfo.msg_perm.uid=8;
msg_sinfo.msg_perm.gid=8;
msg_sinfo.msg_qbytes=16388;
reval=msgctl(msgid,IPC_SET,&msg_sinfo);
if(reval==-1)
{
    printf("msg set info error\n");
    return;
}
msg_stat(msgid,msg_ginfo);
reval=msgctl(msgid,IPC_RMID,NULL);
if(reval==-1)
{
    printf("unlink msg queue error\n");
    return;
```

```c
    }
  }
void msg_stat(int msgid,struct msqid_ds msg_info)
{
   int reval;
   sleep(1);
   reval=msgctl(msgid,IPC_STAT,&msg_info);
   if(reval==-1)
   {
      printf("get msg info error\n");
      return;
   }
   printf("\n");
   printf("current number of bytes on queue is %d\n",msg_info.msg_cbytes);
   printf("number of messages in queue is %d\n",msg_info.msg_qnum);
   printf("max number of bytes on queue is %d\n",msg_info.msg_qbytes);
   printf("pid of last msgsnd is %d\n",msg_info.msg_lspid);
   printf("pid of last msgrcv is %d\n",msg_info.msg_lrpid);
   printf("last msgsnd time is %s", ctime(&(msg_info.msg_stime)));
   printf("last msgrcv time is %s", ctime(&(msg_info.msg_rtime)));
   printf("last change time is %s", ctime(&(msg_info.msg_ctime)));
   printf("msg uid is %d\n",msg_info.msg_perm.uid);
   printf("msg gid is %d\n",msg_info.msg_perm.gid);
}
```

实例 4 通过共享内存的办法，在父进程中输入变量 X，然后将其传递给子进程，在子进程中输入密码，并通过 X 进行简单加密，再将加密后的密码传递给父进程输出。源代码：

```c
#include<sys/types.h>
#include<unistd.h>
#include<sys/mman.h>
#include<fcntl.h>
typedef struct
  {
   int Exchange;
   int  x;
  }code;
main(int argc, char** argv)
{
  pid_t result;
```

```
        code *e_map;
        int y;
    e_map=(code*)mmap(NULL,sizeof(code)*10,PROT_READ|PROT_WRITE,MAP_SHARED|
MAP_ANONYMOUS,-1,0);
        result=fork();
        if(result<0)
            {perror("failed");
             exit(0);
             }
        else if(result==0)
            {sleep(5);
             printf("\nHere is the child process!");
             printf("\nNow you can enter code:");
             scanf("%d",&(*(e_map)).Exchange);
             (*(e_map)).Exchange=(*(e_map)).Exchange*(((*e_map).x))*3+11;
             munmap(e_map,sizeof(code)*10);
             exit(0);
             }
        else
            {printf("Here is the father process!");
             printf("\nNow you can enter the X:");
             scanf("%d",&(*(e_map)).x);
             sleep(8);
             printf("\nNow the code is ");
             printf("%d",(*(e_map)).Exchange);
             munmap(e_map,sizeof(code)*10);
             sleep(2);
             printf("\nThanks for use!");
             }
}
```

实例5 编写程序用消息队列实现服务进程与客户进程的通信。利用消息队列模拟了客户端与服务器之间的交互过程,服务进程首先建立一个消息队列,并读取其中由客户端发来的消息(一个文件的路径名),然后按指定的路径名去读文件,然后将文件内容写回队列。客户端读取服务进程为它返回的文件内容,然后打印。

客户端程序代码:
```
#include <stdio.h>
#include <string.h>
#include <stdlib.h>
#include <sys/types.h>
```

```c
#include <sys/msg.h>
#include <sys/ipc.h>
#include <unistd.h>
struct msgmbuf                      /*结构体,定义消息的结构*/
 {
    long msg_type;                  /*消息类型*/
    char msg_text[512];             /*消息内容*/
 };
int main()
{
int qid;
int len;
struct msgmbuf rmsg,smsg;           /*分别用于储存接受到的消息和将发送的消息*/
qid=msgget(1234,0);                 /*打开消息队列*/
printf("请输入文件的路径:");
if((fgets((&smsg)->msg_text,512,stdin))==NULL)
/*输入的路径存入变量msg_text*/
{
    puts("没有消息");
    exit(1);
}
smsg.msg_type=getpid();
len=strlen(smsg.msg_text);
if((msgsnd(qid,&smsg,len,0))<0)     /*调用msgsnd函数,添加消息到消息队列*/
{
    perror("添加消息出错");
    exit(1);
}
printf("添加路径名成功\n");
if((msgrcv(qid,&rmsg,512,0,0))<0)   /*调用msgrcv函数,从消息队列读取消息*/
{
    perror("读取消息出错");
    exit(1);
}
printf("读取的文件内容是:\n%s\n",(&rmsg)->msg_text);    /*打印文件内容*/
exit(0);
}
```

服务器端程序代码:
```c
#include <stdio.h>
```

```c
#include <string.h>
#include <stdlib.h>
#include <sys/types.h>
#include <sys/msg.h>
#include <sys/ipc.h>
#include <unistd.h>
struct msgmbuf
{
long msg_type;
char msg_text[512];
};
int main()
{
int qid,pid,i=0;
int len;
struct msgmbuf rmsg,smsg;
FILE *fp;
char c;
if((qid=msgget(1234,IPC_CREAT|0666))==-1)
/*调用msgget函数,创建、打开消息队列。简单起见,没有调用ftok产生标准key*/
{
    perror("创建消息队列出错");
    exit(1);
}
printf("创建、打开消息队列(%d)成功\n",qid);
if((msgrcv(qid,&rmsg,512,0,0))<0)
/*调用msgrcv函数,从消息队列读取消息*/
{
    perror("读取路径名出错");
    exit(1);
}
printf("读取路径名成功\n");
pid=rmsg.msg_type;
len=strlen(rmsg.msg_text);
rmsg.msg_text[len-1]=0;         /*将路径名后的回车替换成字符串结束符'\0'*/
if((fp=fopen(rmsg.msg_text,"r"))==NULL)
{
    printf("打开文件出错");
    exit(1);
```

```c
    }
    printf("打开文件成功\n");
    while((c=fgetc(fp))!=EOF)
    {
        smsg.msg_text[i]=c;
        i++;
    }              /*读取文件内容储存到 smsg.msg_text[]*/
    smsg.msg_text[i]=0;
    len=strlen(smsg.msg_text);
    smsg.msg_type=pid;
    if((msgsnd(qid,&smsg,len,0))<0)    /*调用 msgsnd 函数，添加消息到消息队列*/
    {
        perror("添加消息出错");
        exit(1);
    }
    printf("成功返回文件内容\n");
    sleep(1);   /*保证客户端有足够的时间获取文件内容*/
    if((msgctl(qid,IPC_RMID,NULL))<0)
    /*调用 msgctl 函数，删除系统中的消息队列*/
    {
        perror("删除消息队列出错");
        exit(1);
    }
    exit (0);
}
```

实例 6 共享内存程序设计。设计两个.c 文件，要求用系统 V 共享内存通信，在程序中应用 2 个进程分别控制输入输出，达到即时通讯的效果，实现简单的聊天功能。聊天时要求显示系统时间，当输入"QUIT"时退出程序。

Share1.c 源程序：

```c
#include <sys/ipc.h>
#include <sys/shm.h>
#include <sys/types.h>  /*文件预处理，包含 waitpid、kill、raise 等函数库*/
#include <unistd.h>     /*文件预处理，包含进程控制函数库*/
#include <stdlib.h>
#include <stdio.h>
#include <string.h>
#include <time.h>
#define NAMESIZE 60
#define MAXSIZE 512
```

```c
#define QUIT "quit"

typedef struct
{
    int msg_type;
    char msg_user[NAMESIZE];
    char msg_text[MAXSIZE];
    char msg_time[30];
}mymsg;

main(int argc, char** argv)
{
 int shm_id,i;
 key_t key;
 mymsg *msg, *tmpmsg;
 char* name = "/dev/shm/myshm2";
 char usrname[NAMESIZE];
 time_t t;
 pid_t pid;
 key = ftok(name,0); /*调用ftok函数，产生标准的key*/
 shm_id = shmget(key,4096,IPC_CREAT); /*调用shmget函数，获取共享内存区域的ID*/
 if(shm_id == -1)
 {
    perror("ID fail");
    return;
 }
 msg = (mymsg*)shmat(shm_id,NULL,0);
 tmpmsg = msg + 1;
 printf("Please enter a user name: ");
 scanf("%s", usrname);
 printf("%s:",usrname);
 msg->msg_type == 1;
 if((pid=fork())<0)
 {
    printf("error");
    exit(0);
 }
    if(pid==0)
    {
```

```c
        while(1)
          {   if(msg->msg_type == 1)
            {
             printf("\n");
             printf("%s    :%s%s\n    %s:\n",  msg->msg_user,   msg->msg_time, msg->msg_text,usrname);
          msg->msg_type = 0;
            }
    }
    }
    else{
    while(1)
    {
    printf("%s:",usrname);
    fgets(tmpmsg->msg_text,100,stdin);
    if(strcmp(QUIT, tmpmsg->msg_text) == 0)
         {
          printf(" you are quiting now...\n");
           exit(1);
           }
      strcpy(tmpmsg->msg_user, usrname);
      time(&t);
      strcpy(tmpmsg->msg_time, ctime(&t));
      tmpmsg->msg_type = 1;
      }
      }
         if(shmdt(msg) == -1)  /*调用 shmdt 函数,解除进程对共享内存区域的映射*/
         perror("release fail\n");

}
```

Share2.c 源程序:
```c
#include <sys/ipc.h>
#include <sys/shm.h>
#include <sys/types.h>  /*文件预处理,包含 waitpid、kill、raise 等函数库*/
#include <unistd.h>       /*文件预处理,包含进程控制函数库*/
#include <stdlib.h>
#include <stdio.h>
#include <string.h>
#include <time.h>
```

```c
#define NAMESIZE 60
#define MAXSIZE 512
#define QUIT "quit"
typedef struct
{
int msg_type;
char msg_user[NAMESIZE];
char msg_text[MAXSIZE];
char msg_time[30];
}mymsg;
main(int argc, char** argv)
{
 int shm_id,i;
 key_t key;
 char temp;
 mymsg *msg, *tmpmsg;
 char* name = "/dev/shm/myshm2";
 char usrname[NAMESIZE];
 pid_t pid;
 time_t t;
 key = ftok(name,0);                      /*调用ftok函数,产生标准的key*/
 shm_id=shmget(key,4096,IPC_CREAT);       /*调用shmget函数,获取共享内存区域的ID*/
 if(shm_id==-1)
    {
    perror("ID fail");
    return;
    }
    msg = (mymsg*)shmat(shm_id,NULL,0);
tmpmsg = msg + 1;
printf("Please enter a user name: ");
scanf("%s", usrname);
printf("%s:",usrname);
tmpmsg->msg_type=1;
if((pid=fork())<0)
{
        printf("error");
        exit(0);
}
if(pid==0)
```

```
    {
        while(1)
        {
         if(tmpmsg->msg_type == 1)
         {
          printf("\n");

          printf("%s : %s%s\n""%s:\n", tmpmsg->msg_user, tmpmsg->msg_time,
                 tmpmsg->msg_text,usrname);
         tmpmsg->msg_type = 0;
         }
        }
    }
    else{
     while(1)
     {
       printf("%s:",usrname);
       fgets(msg->msg_text,100,stdin);
       if(strcmp(QUIT, msg->msg_text) == 0)
       {
        printf(" you are quiting now...\n");
        break;
        }
        strcpy(msg->msg_user, usrname);
        time(&t);
        strcpy(msg->msg_time, ctime(&t));
        msg->msg_type = 1;
       }
      }
     if(shmdt(msg)==-1)         /*调用 shmdt 函数，解除进程对共享内存区域的映射*/
      perror("release fail\n");
    }
```

第8章 线 程

8.1 知识要点

8.1.1 线程与进程

简单地说，进程是资源管理的最小单位，线程是程序执行的最小单位。一个进程至少需要一个线程作为它的指令执行体，进程管理着资源（比如 CPU、内存、文件等），而将线程分配到某个 CPU 上执行。在操作系统设计上，从进程演化出线程，最主要的目的为：

1. 更好地支持对称多处理机（SMP：Symmetrical Multi-Processing）

一个进程当然可以拥有多个线程，此时，如果进程运行在 SMP 机器上，它就可以同时使用多个 CPU 来执行各个线程，达到最大程度的并行，以提高效率。

2. 减小（进程/线程）上下文切换开销

即使是在单 CPU 的机器上，采用多线程模型来设计程序，使设计更简洁、功能更完备，程序的执行效率也更高，例如采用多个线程响应多个输入。此多线程模型所实现的功能实际上也可以用多进程模型来实现，但与后者相比，线程的上下文切换开销就比进程要小多了。从语义上来说，同时响应多个输入这样的功能，实际上就是共享了除 CPU 以外的所有资源。

Linux 内核只提供了轻量进程的支持，限制了更高效的线程模型的实现，但 Linux 着重优化了进程的调度开销，一定程度上也弥补了这一缺陷。目前最流行的线程机制 Linux

Threads 所采用的就是线程-进程"一对一"模型，调度交给核心，而在用户级实现一个包括信号处理在内的线程管理机制。

8.1.2 多线程和多进程的对比

多线程和多进程在程序设计中没有绝对的好与坏，只有哪个更合适的问题。要看实际应用中究竟如何判断更加合适。

1.需要频繁创建销毁的优先用线程

这种原则最常见的应用就是 Web 服务器了，来一个连接建立一个线程，断了就销毁线程。要是用进程，创建和销毁的代价是很难承受的

2.需要进行大量计算的优先使用线程

所谓大量计算，当然就是要耗费很多 CPU，切换频繁了，这种情况下线程是最合适的。这种情形最常见的是图像处理、算法处理。

3.强相关的处理用线程，弱相关的处理用进程

一般的 Server 需要完成如下任务：消息收发、消息处理。"消息收发"和"消息处理"就是弱相关的任务，而"消息处理"里面可能又分为"消息解码"、"业务处理"，这两个任务相对来说相关性就要强多了。因此"消息收发"和"消息处理"可以分进程设计，"消息解码"、"业务处理"可以分线程设计。

当然这种划分方式不是一成不变的，也可以根据实际情况进行调整。

4.可能要扩展到多机分布的用进程，多核分布的用线程

5.都满足需求的情况下，用最熟悉、最拿手的方式

至于"数据共享、同步"、"编程、调试"、"可靠性"这几个维度的所谓的"复杂、简单"应该怎么取舍，没有明确的选择方法。

8.1.3 线程中的常用函数

（1）线程中的函数 pthread_create 类似于进程中的 fork 函数；

（2）线程中的函数 pthread_join 类似于进程中的 waitpid 函数，等待指定的线程结束；

（3）线程中的函数 pthread_self 类似于进程中的 getpid 函数；

（4）线程中的函数 ptrhead_detach 用于改变线程状态。在 Linux 中线程 pthread 有两种状态 joinable 状态和 unjoinable 状态。如果线程是 joinable 状态，当线程函数自己返回或退出时或 pthread_exit 时都不会释放线程所占用堆栈和线程描述符，只有当你调用了 pthread_join 之后这些资源才会被释放。若是 unjoinable 状态的线程，这些资源在线程函数退出时或调用函数 pthread_exit 时自动会被释放。通过调用函数 pthread_detach(pthread_self())，将状态改为 unjoinable 状态，确保资源的释放。

（5）函数 pthread_exit 使主线程自身退出。在主线程中使用 pthread-exit 函数只会使主线程自身退出，产生的子线程继续执行，用 return 则所有线程退出。

8.1.4 线程中互斥锁的实现

线程最大的特点是资源的共享性，然而资源共享中的同步问题是多线程编程的难点。Linux 提供多种方式处理线程间同步问题包含互斥锁、条件变量、异步信号；在多线程处理同一部分数据的时候，就需要保证数据的一致性。互斥锁的操作主要包括以下几个步骤：

（1）互斥锁初始化：pthread_mutex_init；
（2）互斥锁上锁：pthread_mutex_lock；
（3）互斥锁判断上锁：pthread_mutex_trylock；
（4）互斥锁解锁：pthread_mutex_unlock；
（5）消除互斥锁：pthread_mutex_destroy。

设置一个 pthread_mutex_t 型的共享变量，这个变量的功能就是用来做标记的，一旦这个变量被任何一个线程调用 pthread_mutex_lock()函数标记，那就代表它或者其他可能的共享变量被该线程占用，而其他使用 pthread_mutex_lock()函数的线程就会暂停下来，直到第一个调用 pthread_mutex_lock()对共享变量标记的线程调用 pthread_mutex_unlock()对共享变量解锁，其他要标记 pthread_mutex_t 型共享变量的线程才会继续运行。

例如：

```
pthread_mutex_t cat;
int dog=0;
void * pthread1(void * argu)
{
  pthread_mutex_lock(&cat);
  for (;dog<10; dog++) printf("pthread1: dog=%d\n", dog);
  pthread_mutex_unlockl(&cat);
}
void * pthread2(void * argu)
{
  pthread_mutex_lock(&cat);
  for (;dog<20; dog++) printf("pthread2: dog=%d\n", dog);
  pthread_mutex_unlockl(&cat);
}
int main()
{
  pthread_t tid1, tid2;
  pthread_mutex_init(&cat, NULL);
  pthread_create(&tid1, NULL, (void *)pthread1, NULL);
  pthread_create(&tid1, NULL, (void *)pthread2, NULL);
  pthread_mutex_destroy(&cat);
```

```
    return 0;
}
```
这样就可以保证线程 pthread 1 和 pthread 2 不会同时访问共享变量 dog 了。

8.1.5 线程中信号量的应用

信号量（semaphore）是一个计数器，用于多进程对共享数据对象的访问。为了获得共享资源，进程需要①测试控制该资源的信号量；②若此信号量的值为正，则进程可以使用该资源。进程将信号量减 1，表示它使用了一个资源单位；③若此信号量的值为 0，则进程进入休眠状态，直至信号量值大于 0，进程被唤醒后，返回第 1）步。当进程不再使用由一个信号量控制的共享资源时，该信号量增 1。如果有进程正在休眠等待此信号量，则唤醒它们。常用的信号量形式被称为二元信号量，它控制单个资源，初始值为 1。

1．创建信号量集，调用 semget 函数
```
#include <sys/sem.h>
int semget(key_t key, int nsems, int flag);
```
key 是 IPC_PRIVATE，或者 key 当前未与信号量相结合且 flag 中指定了 IPC_CREAT 位。新信号量集的访问权限根据 flag 中的访问权限位设置。nsems 参数是该集合中的信号量数。如果是创建新集合，则必须指定 nsems，如果引用一个现存的集合，则将 nsems 指定为 0。若执行成功，semget 函数将返回信号量集 ID，用于其它操作。

2．操作信号量集，调用 semctl 函数
```
#include <sys/sem.h>
int semctl(int semid, int semnum, int cmd, ... /* union semun arg */);
```
第四个参数是可选的，cmd 参数有 10 种选择，在 semid 指定的信号量集上执行此命令。其中有 5 条命令是针对一个特定的信号量，它们用 semnum 指定该信号量集中的一个信号量成员，semnum 值在 0 和 nsems-1 之间。

3．获取和释放信号量
获取和释放信号量通过 semop 函数完成，它是一个原子操作。
```
#include <sys/sem.h>
int semop(int semid, struct sembuf semoparray[], size_t nops);
```
参数 semoparray 是一个指针，它指向一个信号量操作数组，信号量操作由 sembuf 结构表示。参数 nops 规定该操作数组的元素个数。对信号量集中成员 sem_num 的操作由相应的 sem_op 值规定。

8.2 程序设计实例

实例 1 Linux 中的临界区域问题。在内存访问时，由于 CPU 的多线程调度缘故，可能出现多个线程对内存中某个共享变量的操作顺序与程序设计时的意图相违背，最终得到意料之外的结果，这一问题即为操作系统中经典的临界区问题。在 Linux 中，可以使用互斥锁来解决这一问题。当需要使用某一临界区时，可以给其加上一个互斥锁，在此期间其他程序即使获得了 CPU 的运行权，也不可访问此加锁的区域，直到使用结束，释放该区域后其他进程才能使用该区域。下列程序中的两个进程均要访问共享变量 counter，使用一个互斥锁 mux，在访问时将其锁起。

```c
#include <stdio.h>
#include <stdlib.h>
#include <unistd.h>
#include <pthread.h>
int counter=0; //the sharing argument
pthread_mutex_t mux; //a mutex lock of <counter>
void thread1(void *arg);
void thread2(void *arg);
int main(int argc, char *argv[])
{
pthread_t id1,id2;
unsigned int slp1,slp2;
/*initial the mutex lock*/
//pthread_mutex_init(&mux,NULL);
/*user can define the sleep time*/
printf("please printf the sleep time slp1 and slp2:\n");
scanf("%d",&slp1);
scanf("%d",&slp2);
/*create 1st thread*/
pthread_create(&id1,NULL,(void *)thread1, (void *)&slp1);
/*create 2st thread*/
pthread_create(&id2,NULL,(void *)thread2, (void *)&slp2);
/*waiting for the two threads completed*/
pthread_join(id1,NULL);
pthread_join(id2,NULL);
/*print the last value of <counter>*/
```

```c
    printf("最后的 counter 值为%d\n",counter);
    exit(0);
}
void thread1(void *arg)   /*第1个线程执行代码*/
{
    int i,val;
    for(i=1;i<=5;i++){
    /*lock*/
    pthread_mutex_lock(&mux);
    val=++counter;
    printf("第1 个线程:第%d 次循环,第1 次引用 counter=%d\n",i,counter);
    /*LINE A*/
    sleep(*((unsigned int *)arg));   /*睡眠或挂起<slp1>秒钟*/
    counter=val;
    /*unlock*/
    pthread_mutex_unlock(&mux);
    }
}
void thread2(void *arg)   /*第2个线程执行代码*/
{
    int i,val;
    for(i=1;i<=5;i++){
    /*lock*/
    pthread_mutex_lock(&mux);
    val=++counter;
    sleep(*((unsigned int *)arg));   /*睡眠或挂起<slp2>秒钟*/
    printf("第2 个线程第%d 次循环,counter=%d\n",i,counter);
    counter=val;
    /*unlock*/
    pthread_mutex_unlock(&mux);
    }
}
```

实例2 编写一个多线程程序来生成 Fibonacci 序列。程序中输入要产生 Fibonacci 序列的个数,然后程序创建一个新的线程来产生 Fibonacci 序列,把这个序列放到线程共享的数据中。当线程执行完成后,父线程将输出由子线程产生的序列。

```c
#include <stdio.h>
#include <fcntl.h>
#include <string.h>
#include <stdlib.h>
```

```c
#include <sys/select.h>
#include <sys/types.h>
#include <sys/stat.h>
#include <errno.h>
#include <signal.h>

int main()
{
    int status; //接收子进程结束时的信息
    pid_t child1, child2; //两个子进程的标识
    int rfd; //管道读标识
    int wfd; //管道写标识
    int *fnum; //存储 F 数的动态分配数组
    int num; //接受用户要求的 F 数
    int i;      //循环变量
    int temp; //临时存放 F 数
    int check; //检查只读管道是否还有数据
    child1 = fork();   //子进程 P1
    if(child1 >= 0)    //进程创建成功
    {
        if(child1 == 0)   //子进程内
        {
            printf("child1 process! %d\n", getpid());   //打印子进程的进程号

        }
        else //父进程内
        {
            printf("parent process!\n");
            wait(&status);    //等待子进程结束
            printf("child1's exit code is %d\n", WEXITSTATUS(status));
        }
    }
    else //进程创建失败
    {
        printf("fork failed!");
        exit(0);
    }

    printf("to start child2\n");
```

```c
child2 = fork();  //创建进程2
if(child2 >= 0)  //进程创建成功
{
    if(child2 == 0)   //子进程内
    {
        printf("child2 process! %d\n", getpid());  //输出子进程号
        mkfifo("fifo1", S_IWUSR|S_IRUSR|S_IRGRP|S_IROTH);  //创建管道
        wfd = open("fifo1", O_WRONLY);  //只写打开管道
        if(wfd <= 0)          //管道打开失败
        {
            printf("wfd error!\n");
            return 0;
        }
        printf("child2 pipe successful!\n");
        printf("How many F numbers do you like to get?\n");
        scanf("%d", &num);  //接受用户要求的F数
        if(num <= 0)
        {
            perror("num less than 1, input wrong!");
        }
        fnum = (int *)malloc(sizeof(int) * num);  //存储F数的动态分配数组
        if(num > 0)
        {
            fnum[0] = 0;  //初始值
            write(wfd, &fnum[0], sizeof(int));  //先将初始值写入管道
        }
        if(num > 1)
        {
            fnum[1] = 1;
            write(wfd, &fnum[1], sizeof(int));
        }

        for(i = 2; i < num; i++)   //根据f数的生成方法生成f数,并逐个写入管道
        {
            fnum[i] = fnum[i - 1] + fnum[i - 2];
            temp = fnum[i];
            write(wfd, &temp, sizeof(int));
```

```
            }
            close(wfd); //关闭管道

            exit(0);
        }
        else    //父进程
        {
            printf("parent process!\n");
            mkfifo("fifo1", S_IWUSR|S_IRUSR|S_IRGRP|S_IROTH); //创建相同的管道
            rfd = open("fifo1", O_RDONLY); //只读打开管道
            if(rfd <= 0)            //管道打开失败
            {
                printf("pipe open failed\n");
                return 0;
            }
            printf("parent pipe successful!\n");
            wait(&status); //等待子进程结束
            printf("child2's exit code is %d\n", WEXITSTATUS(status));
            do
            {
                check = read(rfd, &temp, sizeof(int)); //判断管道中是否还有数据
                if(check != 0)
                    printf("%d\n", temp); //若有数据，打印输出
            }while(check != 0);

            close(rfd); //关闭管道
        }
    }
    else //进程创建失败
    {
        printf("fork failed!");
        exit(0);
    }
```

实例 3　一组生产者线程与一组消费者线程通过缓冲区发生联系，生产者线程将生产的产品送入缓冲区，消费者线程则从中取出产品。缓冲区有 *n* 个，是一个环形的缓冲池。这是以著名的生产者/消费者问题为例来阐述 **Linux** 线程的控制和通信，涉及线程互斥与线程同步概念。

```
#include <stdio.h>
```

```c
#include <pthread.h>
#define BUFFER_SIZE 16 // 缓冲区数量
 // 缓冲区相关数据结构
struct prodcons
{
  int buffer[BUFFER_SIZE]; /* 实际数据存放的数组*/
  pthread_mutex_t lock; /* 互斥体 lock 用于对缓冲区的互斥操作 */
  int readpos, writepos; /* 读写指针*/
  pthread_cond_t notempty; /* 缓冲区非空的条件变量 */
  pthread_cond_t notfull; /* 缓冲区未满的条件变量 */
};

/* 初始化缓冲区结构 */
void init(struct prodcons *b)
{
  pthread_mutex_init(&b->lock, NULL);
  pthread_cond_init(&b->notempty, NULL);
  pthread_cond_init(&b->notfull, NULL);
  b->readpos = 0;
  b->writepos = 0;
}

/* 将产品放入缓冲区，这里是存入一个整数*/
void put(struct prodcons *b, int data)
{
  pthread_mutex_lock(&b->lock);
  /* 等待缓冲区未满*/
  if ((b->writepos + 1) % BUFFER_SIZE == b->readpos)
  {
    pthread_cond_wait(&b->notfull, &b->lock);
  }
  /* 写数据，并移动指针 */
  b->buffer[b->writepos] = data;
  b->writepos++;
  if (b->writepos >= BUFFER_SIZE)
    b->writepos = 0;
  /* 设置缓冲区非空的条件变量*/
  pthread_cond_signal(&b->notempty);
  pthread_mutex_unlock(&b->lock);
```

```c
}

/* 从缓冲区中取出整数*/
int get(struct prodcons *b)
{
  int data;
  pthread_mutex_lock(&b->lock);
  /* 等待缓冲区非空*/
  if (b->writepos == b->readpos)
  {
    pthread_cond_wait(&b->notempty, &b->lock);
  }
  /* 读数据，移动读指针*/
  data = b->buffer[b->readpos];
  b->readpos++;
  if (b->readpos > = BUFFER_SIZE)
    b->readpos = 0;
  /* 设置缓冲区未满的条件变量*/
  pthread_cond_signal(&b->notfull);
  pthread_mutex_unlock(&b->lock);
  return data;
}

/* 测试：生产者线程将1 到10000 的整数送入缓冲区，消费者线
程从缓冲区中获取整数，两者都打印信息*/
#define OVER ( - 1)
struct prodcons buffer;
void *producer(void *data)
{
  int n;
  for (n = 0; n < 10000; n++)
  {
    printf("%d --->"n", n);
    put(&buffer, n);
  }
  put(&buffer, OVER);
  return NULL;
}
```

```c
void *consumer(void *data)
{
  int d;
  while (1)
  {
    d = get(&buffer);
    if (d == OVER)
      break;
    printf("--->%d "n", d);
  }
  return NULL;
}

int main(void)
{
  pthread_t th_a, th_b;
  void *retval;
  init(&buffer);
  /* 创建生产者和消费者线程*/
  pthread_create(&th_a, NULL, producer, 0);
  pthread_create(&th_b, NULL, consumer, 0);
  /* 等待两个线程结束*/
  pthread_join(th_a, &retval);
  pthread_join(th_b, &retval);
  return 0;
}
```

实例4　通过创建两个线程来实现对一个数的递加，分别在各自线程完成1→10之间的加法，程序代码如下：

```c
#include <pthread.h>
#include <stdio.h>
#include <sys/time.h>
#include <string.h>
#define MAX 10
pthread_t thread[2];
pthread_mutex_t mut;
int number=0, i;
void *thread1()
{
        printf ("thread1 : I'm thread 1\n");
```

```c
        for (i = 0; i < MAX; i++)
        {
                printf("thread1 : number = %d\n",number);
                pthread_mutex_lock(&mut);
                number++;
                pthread_mutex_unlock(&mut);
                sleep(2);
        }
        printf("thread1 :主函数在等我完成任务吗？\n");
        pthread_exit(NULL);
}
void *thread2()
{
        printf("thread2 : I'm thread 2\n");
        for (i = 0; i < MAX; i++)
        {
                printf("thread2 : number = %d\n",number);
                pthread_mutex_lock(&mut);
                number++;
                pthread_mutex_unlock(&mut);
                sleep(3);
        }
        printf("thread2 :主函数在等我完成任务吗？\n");
        pthread_exit(NULL);
}
void thread_create(void)
{
        int temp;
        memset(&thread, 0, sizeof(thread));
        if((temp = pthread_create(&thread[0], NULL, thread1, NULL)) != 0)
                printf("线程1创建失败!\n");
        else
                printf("线程1被创建\n");
        if((temp = pthread_create(&thread[1], NULL, thread2, NULL)) != 0)
                printf("线程2创建失败");
        else
                printf("线程2被创建\n");
}
void thread_wait(void)
```

```c
{
    /*等待线程结束*/
    if(thread[0] !=0) {                     //comment4
        pthread_join(thread[0],NULL);
        printf("线程1已经结束\n");
    }
    if(thread[1] !=0) {                     //comment5
        pthread_join(thread[1],NULL);
        printf("线程2已经结束\n");
    }
}
int main()
{
    /*用默认属性初始化互斥锁*/
    pthread_mutex_init(&mut,NULL);
    printf("我是主函数哦，我正在创建线程，呵呵\n");
    thread_create();
    printf("我是主函数哦，我正在等待线程完成任务阿，呵呵\n");
    thread_wait();
    return 0;
}
```

第9章 网络程序设计

9.1 知识要点

9.1.1 socket 接口

socket 是网络编程的一种接口,是一种特殊的 I/O。使用 socket 函数可以建立一个 socket 连接。通常 socket 分为流式 socket、数据报 socket 及原始 socket。socket 根据所使用协议的不同,可分为 TCP 套接口和 UDP 套接口,也分别称为流式套接口和数据套接口。UDP 是无连接协议,传输 UDP 数据包时,系统不关心他们是否安全到达目的地,而传输 TCP 数据包时,应先建立连接,以保证传输数据包被正确接收。

9.1.2 sockaddr 和 sockaddr_in 结构类型

sockaddr 和 sockaddr_in 都是用来保存一个套接字的信息。不同点是 sockaddr_in 将 IP 地址与端口分为不同的成员。sockaddr_in 较 sockaddr 比较方便,可以轻松地处理套接字地址的基本元素。sockaddr 的定义方法为:
```
struct sockaddr
{
```

```
 unsigned short int sa_family;
 char sa_data[14];
};
```
sockaddr_in 的定义方法为:
```
struct sockaddr_in
{
 unsigned short int sin_family;
 unit16_t sin_port;
 struct in_addr sin_addr;
 unsigned char sin_zero[8];
};
```
两个结构体的成员中 sa_family 与 sin_family 都是指定通信的地址类型。

sa_data 用来保存 IP 地址和端口信息,最多 14 个字符长度。

sin_port 是套接字使用的端口号。

sin_addr 是需要访问的 IP 地址(注意 in_addr 也是个结构体)。

sin_zero 未使用的字段,填充为 0。用来将 sockaddr_in 结构填充到与 struct sockaddr 相同的长度,可以用 bzero()函数将其置为零。

注意:

在 socket 编程中想用 struct sockaddr 结构变量时仍可以用 struct sockaddr_in 定义,只要用函数 bzero 将 sockaddr_in 中的 sin_zero 置为 0 就可以了。

例如:
```
my_addr.sin_family=AF_INET;
my_addr.sin_port = htons(SERVPORT);
my_addr.sin_addr.s_addr=INADDR_ANY;
bzero(&(my_addr.sin_zero),8);
```

9.1.3 TCP 编程

1.服务端编程过程

(1) 建立 socket 连接,例:

socket(AF_INET, SOCK_STREAM, 0);

AF_INET 表示进行 IPv4 协议通信,SOCK_STREAM 表示采用流式 socket 即 TCP;

(2) 将 socket 与本机上一个端口绑定,例:

bind(sockfd,(struct sockaddr*)&my_addr, sizeof(struct sockaddr));

可以在该端口监听服务请求,my_addr 包含本机 IP 和端口号的信息;

(3) 建立监听,例:

listen(sockfd, BACKLOG);

使 socket 处于被动的监听模式,并为其建立一个输入数据队列,用来保存到达的服务请求,

BACKLOG 表示最大连接数；

（4）响应客户请求，例：

　accept(sockfd，(struct sockaddr *)&remote_addr，&sin_size);

用函数 accept 生成新的套接口描述符，使服务器接受客户的连续请求，remote_addr 用于接收客户地址信息，sin_size 用于存放地址长度；

（5）面向连接的 socket 发送数据，例：

　send(client_fd, buf, len, 0);

buf 为发送字符串的内存地址，len 表示发送字符串的长度；

（6）停止在该 socket 上的任何操作，例：

　socket close(client_fd);

例如：

```
……
sockfd=socket(AF_INET, SOCK_STREAM, 0);
my_addr.sin_family = AF_INET;
my_addr.sin_port = htons(SERVPORT);
my_addr.sin_addr.s_addr = INADDR_ANY;
bzero(&(my_addr.sin_zero), 8);
bind(sockfd, (struct sockaddr *)&my_addr, sizeof(struct sockaddr));
listen(sockfd, BACKLOG);
client_fd = accept(sockfd, (struct sockaddr *)&remote_addr, &sin_size);
……
```

2.客户端编程过程

（1）建立 socket 连接，例：

　socket(AF_INET，SOCK_STREAM，0);

为相应数据结构分配存储空间；

（2）请求连接，例：

　connect(socket_fd, (struct sockaddr*) &serv_addr, sizeof(struct sockaddr));

启动和远端主机的直接连接；

（3）接收数据，例：

　recv(sockfd, buf, MAXDATASIZE, 0);

（4）关闭 socket, 例：

　close(sockfd);

例如：

```
……
host = gethostbyname("xxx.xxx.xxx.xxx");
sockfd = socket(AF_INET, SOCK_STREAM, 0);
ser_addr.sin_family = AF_INET;
ser_addr.sin_port = htons(SERVPORT);
ser_addr.sin_addr.s_addr =* ((struct in_addr *)host->h_addr);
```

```
bzero(&(ser_addr.sin_zero), 8);
connect(sockfd, (struct sockaddr *)&ser_addr, sizeof(struct sockaddr));
......
```

9.1.4 UDP 编程

UDP 主要特点是客户端不需要使用 bind 函数把本地 IP 地址与端口号进行绑定也能进行通信。

1.服务端

（1）建立 socket 并为其分配空间，例：
　　socket(AF_INET，SOCK_DGRAM，0);
（2）绑定 socket 与本机上的一个端口，可以在该端口监听服务请求，例：
　　bind(sockfd, (struct sockaddr *)&adr_inet, sizeof(adr_inet));
（3）接收数据，recvfrom，用于进行数据接收，例：
　　recvfrom(sockfd, buf, sizeof(bof), 0, (struct sockaddr *)&adr_clnt, &len);
（4）停止在该 socket 上的任何数据操作，例：
　　close(sockfd)。

2.客户端

（1）建立 socket 并为其分配空间，例：
　　socket(AF_INET, SOCK_DGRAM, 0);
（2）发送数据，面向连接的 socket 进行数据传输，例：
　　sendto(sockfd,　buf, sizeof(buf), 0, (struct sockaddr *)&adr_srvr, sizeof(adr_srvr));
（3）停止在该 socket 上的任何数据操作，例：
　　close(sockfd)。

9.1.5 常用操作函数

1.socket

创建 socket 描述符
```
 int socket(int domain, int type, int protocol);
```
domain＝AF_INET，socket 的类型，type＝SOCK_STREAM 或 SOCK_DGRAM，分别表示 TCP 连接或 UDP 连接；protocol＝0。

返回一个整型 socket 描述符。

2.bind

将 socket 描述符与本机上的一个端口相关联（仅用于服务器）
```
 int bind(int sockfd, struct sockaddr *my_addr, int addrlen);
```
sockfd 是一个 socket 描述符。

my_addr 是一个指向包含有本机 IP 地址及端口号等信息的 sockaddr 类型的指针；
```
addrlen=sizeof(struct sockaddr).
```
返回：成功＝0；失败＝-1，errno＝错误号。
可以用下面的赋值自动获得本机 IP 地址和随机获取一个没有被占用的端口号：
```
my_addr.sin_port = 0;
my_addr.sin_addr.s_addr = INADDR_ANY;
```

3.connect

与远端服务器建立一个 TCP 连接（用于客户端）
```
int connect(int sockfd, struct sockaddr *serv_addr, int addrlen);
```
sockfd 是目的服务器的 sockt 描述符。
serv_addr 是包含目的机 IP 地址和端口号的指针。
返回：成功＝0；失败＝-1，errno＝错误号。

4.listen

监听是否有服务请求，用于 bind() 后
```
int listen(int sockfd, int backlog);
```
sockfd 是 socket 系统调用返回的 socket 描述符。
backlog 指定在请求队列中允许的最大请求数，缺省值为 20。
返回：成功＝0；失败＝-1，errno＝错误号。

5.accept

接受来自客户的请求
```
int accept(int sockfd, void *addr, int *addrlen);
```
sockfd 是被监听的 socket 描述符。
addr 是指向 sockaddr_in 变量的指针，存放客户主机的信息。
addrten 指向值为 sizeof(struct sockaddr_in) 的整型指针变量。
返回：成功返回一个新的 socket 描述符，来供这个新连接来使用。错误发生时返回-1 并且设置相应的 errno 值。

6.send

在连接(TCP)的 socket 方式下发送信息
```
int send(int sockfd, const void *msg, int len, int flags);
```
Sockfd 是用来传输数据的 socket 描述符。
msg 是一个指向要发送数据的指针。
len 是以字节为单位的数据的长度。
flags 一般情况下置为 0。

7.recv

在连接(TCP)的 socket 方式下接收数据
```
int recv(int sockfd,void *buf,int len,unsigned int flags);
```
sockfd 是接受数据的 socket 描述符；
buf 是存放接收数据的缓冲区；
len 是缓冲的长度。

flags 也被置为 0。

返回：实际上接收的字节数，如果连接中止，返回 0。出现错误时，返回-1 并置相应的 errno 值。

8.close( )

释放 socket，停止任何数据操作

```
close(sockfd);
```

例如：

```
printf("请输入目标端口：");
scanf("%d",&peerport);
printf("请输入本地端口：");
scanf("%d",&localport);
socklen=sizeof(struct sockaddr_in);
/*设定目标 IP 地址*/
peeraddr.sin_family=AF_INET;
peeraddr.sin_port=htons(peerport);
peeraddr.sin_addr.s_addr=htonl(INADDR_ANY);
bzero(&(peeraddr.sin_zero),8);
/*设定本地 IP 地址*/
localaddr.sin_family=AF_INET;
localaddr.sin_port=htons(localport);
localaddr.sin_addr.s_addr=htonl(INADDR_ANY);
bzero(&(localaddr.sin_zero),8);
/*建立 socket*/
sockfd=socket(AF_INET,SOCK_DGRAM,0);
if(sockfd==-1)
{
printf("socket 出错");
    exit(1);
}
/*绑定 socket*/
if(bind(sockfd,(struct sockaddr *)&localaddr,socklen)<0)
{
printf("bind 出错");
exit(1);
}
printf("可以开始通讯：\n 输入任意字符开始：");
scanf("%s",buf+9);              /*前 9 个单元存放时间和回车*/
gettime(buf);                   /*将时间存入 buf[]的前 9 个单元*/
/*发送时间和输入的字符串*/
```

```
sendto(sockfd,buf,strlen(buf),0,(struct sockaddr *)&peeraddr,socklen);
while(1)        /*无限循环，不断接受和发送，终止程序需按下 Ctrl+C*/
{
n=recvfrom(sockfd,buf,255,0,NULL,&socklen);      /*接收字串*/
buf[n]=0;
printf("Peer  %s\n",buf);
printf("Local\n");
scanf("%s",buf+9);
gettime(buf);
sendto(sockfd,buf,strlen(buf),0,(struct sockaddr *)&peeraddr,socklen);
}
```

9.2　程序设计实例

实例1　网络聊天程序中利用多线程处理阻塞。编写一个基于 TCP/IP 的聊天程序，关键要解决的是阻塞问题。在大多数情况下是应用 fcntl()和 select()函数来进行非阻塞设置，而在本题目中采用多线程的方法来解决阻塞的问题。对于服务端来说，连接成功后，主线程负责读取输入并发送信息到客户端，子线程负责接收从客户端发来的信息并输出到屏幕上，对于客户端来说也相同。程序设计中的难点是要注意各个变量之间的的关系，哪一些是线程共享变量，哪一些是线程独占变量。程序要求①本程序既集成了客户端，也集成了服务端：当没有参数时，程序是服务端；当有作为 IP 地址的参数时，程序是客户端。②程序的服务端运行一开始就会有提示，显示自己服务端主机的 IP 地址，以方便客户端连接。③每次聊天信息都会附加时间，并且退出后会有聊天的记录和退出的记录。④程序使用多线程的方法解决阻塞的问题，使得聊天可以并行操作。程序主要框架为：

```
int sockfd, clientfd;
pthread_recv(void * argu)
{
  char recvstr[200];
  int len;
  while(1)
  {
    len=recv(clientfd, recvstr, 200, 0);
    recvdtr[len] = '\0';
    printf("customer: %s",recv);
  }
}
```

```c
int main()
{
    pthread_t ntid;
    char sendstr[200];
    ......
    pthread_create(&ntid, NULL, pthread_recv, NULL);
    while (1)
    {
        gets(sendstr);
        send(clientfd, sendstr, 200, 0);
    }
......
}
```

程序代码：

```c
#include<stdio.h>
#include<stdlib.h>
#include<string.h>
#include<time.h>
#include<unistd.h>
#include<pthread.h>
#include<sys/types.h>
#include<netinet/in.h>
#include<netdb.h>
#include<sys/socket.h>
#include<sys/wait.h>
#include<errno.h>
#include <arpa/inet.h>
#include <fcntl.h>

#define EXITTOSIG "exit###"
#define SERVPORT 3333
#define BACKLOG 10
#define MAXDATASIZE 256

int sockfd;
int oppofd;
struct hostent * host;
struct sockaddr_in serv_addr;
struct sockaddr_in my_addr;
```

```c
struct sockaddr_in remote_addr;
FILE * fp;
pthread_t ntid;

int sendexit();

void * pthread_recv(void * arg)
{
int recvbytes;
char rstr[MAXDATASIZE];
char timestr[20];

while (1) {
    recvbytes = recv(oppofd, rstr, MAXDATASIZE, 0);
    if (recvbytes == -1) {
        perror("recv failed!\n");
        exit(1);
    }
    else if (strcmp(rstr, EXITTOSIG) == 0) {
        printf("The opposite has exited.\n");
        fclose(fp);
        exit(0);
    }
    rstr[recvbytes] = '\0';

    fprintf(fp, "%s", rstr);
    fprintf(fp,"\n");
    printf("\n%s", rstr);
}
pthread_exit((void *) 1);
}

int main(int argc, char * argv[])
{
int sin_size;
int pth_err;
int recvbytes;
char buf[MAXDATASIZE];
char sstr[MAXDATASIZE];
```

```c
    char writing[MAXDATASIZE];
    char chatname[16];
    time_t timep;

    /*------------------------------------------------------------------*/
    if (argc >= 2) {
    printf("\nWhat's your name?\n");
    gets(chatname);

    if ( (fp = fopen("./record_cus.txt", "a+")) == NULL ) {
        perror("creating fp failed!\n");
        exit(1);
    }

    if ( (host = gethostbyname(argv[1])) == NULL ) {
        perror("gethost error!\n");
        exit(1);
    }

    if ((oppofd = socket(AF_INET, SOCK_STREAM, 0)) == -1) {
        perror("socket error!\n");
        exit(1);
    }

    bzero(&serv_addr, sizeof(struct sockaddr_in));
    serv_addr.sin_family=AF_INET;
    serv_addr.sin_port=htons(SERVPORT);
    serv_addr.sin_addr = *((struct in_addr *)host->h_addr);

    if (connect(oppofd, (struct sockaddr *)&serv_addr, sizeof(struct sockaddr))
== -1) {
        perror("connect error!\n");
        exit(1);
    }
    if ( (recvbytes = recv(oppofd, buf, MAXDATASIZE, 0)) == -1 ) {
        perror("received error!\n");
        exit(1);
    }
    buf[recvbytes] = '\0';
```

```c
        printf("the first received information is :%s\n", buf);


        pth_err = pthread_create(&ntid, NULL, pthread_recv, NULL);

        while (1) {
            gets(writing);

            if (strcmp(writing, "exit") == 0) {
                sendexit();
                exit(0);
            }

            strcpy(sstr, chatname);
            strcat(sstr, writing);
            strcat(sstr, "\n");
            strcat(sstr, " at ");
            time(&timep);
            strcat(sstr, asctime(gmtime(&timep)));
            strcat(sstr, "\n");

            fprintf(fp, "%s", sstr);
            fprintf(fp, " \n");
            printf("%s", sstr);
            if ( send(oppofd, sstr, strlen(sstr), 0) == -1 ) {
                perror("send sstr failed!\n");
                exit(1);
            }

            memset(sstr, 0, MAXDATASIZE);
            memset(writing, 0, MAXDATASIZE);
        }

    close(oppofd);
    fclose(fp);
    }
    /*----------------------------------------------------------------*/
    else {
    if ( (fp = fopen("./record_ser.txt", "a+")) == NULL ) {
        perror("creating fp failed!\n");
```

```c
        exit(1);
    }

    if ( (sockfd = socket(AF_INET, SOCK_STREAM, 0)) == -1 ) {
        perror("socket failed!\n");
        exit(1);
    }
    bzero(&my_addr, sizeof(struct sockaddr_in));
    my_addr.sin_family=AF_INET; /*地址族*/
    my_addr.sin_port=htons(SERVPORT);
    my_addr.sin_addr.s_addr = INADDR_ANY;

    if ( bind(sockfd,(struct sockaddr *)&my_addr,sizeof(struct sockaddr))==-1)
    {
        perror("bind failed!\n");
        exit(1);
    }

    printf("Server's IP:\n");
    system("ifconfig | grep inet\\ addr:");
    printf("\n");

    printf("\nWhat's your name?\n");
    gets(chatname);

    if ( listen(sockfd, BACKLOG) == -1 ) {
        perror("listen failed!\n");
        exit(1);
    }

    sin_size = sizeof(struct sockaddr_in);

    while(1) {
        oppofd = accept(sockfd, (struct sockaddr *)&remote_addr, &sin_size);
        if ( oppofd == -1 ) {
            perror("accept failed!\n");
            continue;
        }
        else break;
```

```c
}

printf("received a connection of %s\n", inet_ntoa(remote_addr.sin_addr));
if ( send(oppofd, "sucessfully connceted!\n", 26, 0) == -1 ) {
    perror("first sending failed!\n");
    exit(1);
}

pth_err = pthread_create(&ntid, NULL, pthread_recv, NULL);
if (pth_err != 0) {
    perror("pthread failed!\n");
    exit(1);
}

while (1) {     /*读取输入并发送*/
    gets(writing);
    if (strcmp(writing, "exit") == 0) {
        sendexit();
        exit(0);
    }

    strcpy(sstr, chatname);
    strcat(sstr, writing);
    strcat(sstr, "\n");
    strcat(sstr, " at ");
    time(&timep);
    strcat(sstr, asctime(gmtime(&timep)));
    strcat(sstr, "\n");

    fprintf(fp, "%s", sstr);
    fprintf(fp, "  \n");
    printf("\n%s", sstr);
    if ( send(oppofd, sstr, strlen(sstr), 0) == -1 ) {
        perror("send sstr failed!\n");
        exit(1);
    }

    memset(sstr, 0, MAXDATASIZE);
    memset(writing, 0, MAXDATASIZE);
```

}

```
    close(oppofd);
    close(sockfd);
    fclose(fp);
```

}

/*--*/
```
    return 0;
}

int sendexit()
{
send(oppofd, EXITTOSIG, strlen(EXITTOSIG), 0);
return 0;
}
```

实例 2 设计一个网络聊天程序,服务器端在每次登录时可设置密码一次,客户端需要输入密码才可以成功连接;在聊天时,若某一方输入以 exit 开头的内容,可询问是否要退出聊天程序。

服务器端:

```c
#include <stdlib.h>
#include <stdio.h>
#include <netdb.h>
#include <sys/types.h>
#include <sys/socket.h>
#include <string.h>
#include <netinet/in.h>
#include <arpa/inet.h>
#include <unistd.h>
#include <fcntl.h>
#define MAXDATASIZE 256
#define SERVPORT 4444                    /*监听端口号*/
#define BACKLOG 10                       /*最大同时连接请求数*/
#define STDIN 0
int main(void)
{
FILE *fp;
int sockfd,client_fd;
```

```c
        int sin_size;
        struct sockaddr_in my_addr,remote_addr;
        char buf[256];                  /*聊天用*/
        char buff[256];                 /*用户名用*/
        char send_str[256];
        char choice;
        int recvbytes;
        fd_set rfd_set, wfd_set, efd_set;
        struct timeval timeout;         /*select 最大允许时间*/
        int ret;                        /*select 结果*/
        char serv_pwd[10],serv_repwd[10];
        char trial_pwd[10];
        if ((sockfd = socket(AF_INET, SOCK_STREAM, 0)) == -1) {
        perror("socket");
        exit(1);
        }
        bzero(&my_addr, sizeof(struct sockaddr_in));
        my_addr.sin_family=AF_INET;
        my_addr.sin_port=htons(SERVPORT);
        inet_aton("127.0.0.2", &my_addr.sin_addr);
        if (bind(sockfd, (struct sockaddr *)&my_addr, sizeof(struct sockaddr)) == -1)
        {
        perror("bind");
        exit(1);
        }
        /*监听*/
        if (listen(sockfd, BACKLOG) == -1) {
        perror("listen");
        exit(1);
        }
        /*设置连接密码*/
        printf("请先设置连接密码：\n 请输入: ");
        scanf("%s", serv_pwd);
        printf("请确认密码: ");
        scanf("%s", serv_repwd);
        while (strcmp(serv_pwd, serv_repwd)!=0){
        printf("两次输入的密码不一致，请重试！\n 请输入:");
        scanf("%s", serv_pwd);
```

```c
        printf("请确认密码：");
        scanf("%s", serv_repwd);
    }
    printf("密码设置成功！\n");
    /*连接*/
    sin_size = sizeof(struct sockaddr_in);
    if ((client_fd = accept(sockfd, (struct sockaddr *)&remote_addr, &sin_size)) == -1) {
        perror("accept");
        exit(1);
    }
    fcntl(client_fd, F_SETFD, O_NONBLOCK);                  /*服务器设为非阻塞*/
    /*验证密码*/
    recvbytes=recv(client_fd, trial_pwd, MAXDATASIZE, 0);
    trial_pwd[recvbytes]='\0';
    while (strcmp(trial_pwd,serv_pwd)!=0){
        send(client_fd,"登录密码错误，请重试！",sizeof("登录密码错误，请重试！"),0);
        recvbytes=recv(client_fd, trial_pwd, MAXDATASIZE, 0);
        trial_pwd[recvbytes]='\0';
    }
    send(client_fd,"登录成功！",sizeof("登录成功！"),0);
    recvbytes=recv(client_fd, buff, MAXDATASIZE, 0);         /*接收用户名*/
    buff[recvbytes] = '\0';
    printf("用户%s 成功登录！\n",buff);
    fflush(stdout);                                          /*清除 stdout*/
    if((fp=fopen("name.txt","a+"))==NULL) {
        printf("can not open file,exit...\n");
        return -1;
    }
    fprintf(fp,"%s\n",buff);
    while (1) {
        FD_ZERO(&rfd_set);
        FD_ZERO(&wfd_set);
        FD_ZERO(&efd_set);
        FD_SET(STDIN, &rfd_set);
        FD_SET(client_fd, &rfd_set);
        FD_SET(client_fd, &wfd_set);
        FD_SET(client_fd, &efd_set);
        timeout.tv_sec = 10;
```

```c
timeout.tv_usec = 0;
ret = select(client_fd+1, &rfd_set, &wfd_set, &efd_set, &timeout);
/*若客户端无输入,重新循环*/
if (ret==0) {
continue;
}
if (ret<0) {
perror("select");
exit(-1);
}
//判断标准输入是否有输入
if(FD_ISSET(STDIN,&rfd_set))
{
fgets(send_str, 256, stdin);
send_str[strlen(send_str)-1] = '\0';
//标准输入接收到exit时的处理
if (strncmp("exit",send_str,4)==0) {
printf("确定要退出吗? y or n\n");
scanf("%c",&choice);
if (choice=='y' ||choice=='Y'){
close(client_fd);
close(sockfd);
exit(0);
}
}
//若不退出,照常发送内容给客户端
send(client_fd,send_str,strlen(send_str), 0);
}
//判断客户端是否有输入
if (FD_ISSET(client_fd,&rfd_set)) {
recvbytes=recv(client_fd,buf,MAXDATASIZE, 0);
if (recvbytes==0) {
close(client_fd);
close(sockfd);
exit(0);
}
buf[recvbytes]='\0';
printf("%s:%s\n",buff,buf);
printf("Server: ");
```

```
            fflush(stdout);
        }
        //判断是否有新的描述符加入异常的文件描述符集合
        if (FD_ISSET(client_fd, &efd_set)) {
        close(sockfd);
        close(client_fd);
        exit(0);
        }
    }
}
```

客户端:
```
#include <stdlib.h>
#include <stdio.h>
#include <netdb.h>
#include <sys/types.h>
#include <sys/socket.h>
#include <string.h>
#include <netinet/in.h>
#include <arpa/inet.h>
#include <unistd.h>
#include <fcntl.h>
#define SERVPORT 4444
#define MAXDATASIZE 256
#define STDIN 0

int main()
{
int sockfd;
int recvbytes;
char buf[MAXDATASIZE];
char name[MAXDATASIZE];
char send_str[MAXDATASIZE];
char client_pwd[10],msg[20];
struct sockaddr_in serv_addr;
fd_set rfd_set, wfd_set, efd_set;
struct timeval timeout;
int ret;
if ((sockfd = socket(AF_INET, SOCK_STREAM, 0)) == -1) {
```

```c
    perror("socket 建立失败");
    exit(1);
}
/* 设置 serv_addr */
bzero(&serv_addr, sizeof(struct sockaddr_in));
serv_addr.sin_family=AF_INET;
serv_addr.sin_port=htons(SERVPORT);
inet_aton("127.0.0.2", &serv_addr.sin_addr);
    if (connect(sockfd, (struct sockaddr *)&serv_addr, sizeof(struct sockaddr)) == -1)
{
    perror("connect");
    exit(1);
}
fcntl(sockfd, F_SETFD, O_NONBLOCK);
/*发送密码使服务器检验*/
while (strcmp(msg,"登录成功!")!=0){
printf("请输入登录密码: ");
scanf("%s",client_pwd);
send(sockfd, client_pwd, strlen(client_pwd), 0);
recvbytes=recv(sockfd, msg, MAXDATASIZE, 0);
msg[recvbytes]='\0';
if (strcmp(msg,"登录成功!") != 0) printf("密码错误,请重试!\n");
}
printf("成功登录服务器!\n");
printf("聊天前请输入您的昵称: ");
scanf("%s",name);
name[strlen(name)] = '\0';
printf("%s: ",name);
fflush(stdout);
send(sockfd, name, strlen(name), 0);
while (1)
{
FD_ZERO(&rfd_set);
FD_ZERO(&wfd_set);
FD_ZERO(&efd_set);
FD_SET(STDIN, &rfd_set);
FD_SET(sockfd, &rfd_set);
FD_SET(sockfd, &efd_set);
```

```c
        timeout.tv_sec = 10;
        timeout.tv_usec = 0;
        ret = select(sockfd + 1, &rfd_set, &wfd_set, &efd_set, &timeout);
        if (ret == 0) {             /*若服务器无输入，重新循环*/
        continue;
        }
        if (ret < 0) {
        perror("select error: ");
        exit(-1);
        }
        //判断标准输入是否有输入
        if (FD_ISSET(STDIN, &rfd_set))
        {
        fgets(send_str, 256, stdin);
        send_str[strlen(send_str)-1] = '\0';
        //标准输入接收到 exit 时的处理
        if (strncmp("exit",send_str,4)==0) {
        printf("确定要退出吗? y or n\n");
        scanf("%c",&choice);
        if (choice=='y' ||choice=='Y'){
        close(client_fd);
        close(sockfd);
        exit(0);
        }
        }
        //若不退出，照常发送内容给服务器端
        send(sockfd, send_str, strlen(send_str), 0);
        }
        //判断服务器端是否有输入
        if (FD_ISSET(sockfd, &rfd_set))
        {
        recvbytes=recv(sockfd, buf, MAXDATASIZE, 0);
        if (recvbytes == 0)
        {
        close(sockfd);
        exit(0);
        }
        buf[recvbytes] = '\0';
        printf("Server: %s\n", buf);
```

```
    printf("%s: ",name);
    fflush(stdout);
}
//判断是否有新的描述符加入异常的文件描述符集合
if (FD_ISSET(sockfd, &efd_set))
{
    close(sockfd);
    exit(0);
}
}
    }
```

第 10 章 Linux 图形程序设计

10.1 知识要点

10.1.1 SDL 库

SDL（Simple DirectMedia Layer）是一个跨平台的多媒体游戏支持库。其中包含了对图形、声音、游戏杆、线程等的支持，目前可以运行在许多平台上，其中包括 X Window、X Window with DGA、Linux FrameBuffer 控制台、Linux SVGALib，以及 Windows DirectX、BeOS 等。

因为 SDL 是专门为游戏和多媒体应用而设计开发的，所以它对图形的支持非常优秀，尤其是高级图形能力，比如 Alpha 混和、透明处理、YUV 覆盖、Gamma 校正等。而且在 SDL 环境中能够非常方便地加载支持 OpenGL 的 Mesa 库，从而提供对二维和三维图形的支持。

可以说，SDL 是编写跨平台游戏和多媒体应用程序的最佳平台，并且得到了广泛应用。

10.1.2 常用 SDL 库函数

1.常用的图形模式初始化函数
SDL_Init：加载和初始化 SDL 库；
SDL_SetVideoMode：设置屏幕的视频模式；
SDL_Quit：退出图形模式；
SDL_MapRGB：用像素格式绘制一个 RGB 颜色值；
SDL_FillRect：填充矩形区域；
SDL_UpdateRect：更新指定的区域；
SDL_Delay：延迟一个指定的时间。
例如：
初始化视频子系统，屏幕分辨率设置为 640*480。

```
if(SDL_Init(SDL_INIT_VIDEO)<0)
{   /*初始化视频子系统失败*/
    fprintf(stderr,"无法初始化 SDL: %s\n",SDL_GetError());
    exit(1);
}
screen=SDL_SetVideoMode(640,480,16,SDL_SWSURFACE);   /*设置视频模式*/
```

2.常用的图像函数
SDL_LoadBMP：装载 BMP 位图文件；
SDL_BlitSurface：将图像按设定的方式显示在屏幕上；
SDL_DisplayFormat：图片输出控制；
SDL_FreeSurface：释放图片内存；
TTF_OpenFont：打开字体库，设置字体大小；
TTF_SetFontStyle：设置字体样式；
TTF_RenderUTF8_Blended：渲染文字生成新的 surface；
TTF_Init：初始化 TrueType 字体库；
TTF_CloseFont：释放字体所用的内存空间；
TTF_Quit：关闭 TrueType 字体。
例如：

```
SDL_Surface *LoadImage(char *datafile)            /*装载图像*/
{
SDL_Surface *image,*surface;
image=SDL_LoadBMP(datafile);
surface=SDL_DisplayFormat(image);
SDL_FreeSurface(image);
return(surface);
```

}

例如:

```
void DrawObject(object *obj,SDL_Surface *screen)        /*画目标物件*/
{
SDL_Rect draw;
SDL_Surface *ObjImage;
draw.x=obj->x;
draw.y=obj->y;
draw.w=obj->image->w;
draw.h=obj->image->h;
ObjImage=obj->image;
SDL_BlitSurface(ObjImage,NULL,screen,&draw);            /*显示图像*/
SDL_UpdateRects(screen,1,&draw);                        /*更新图像*/
}
```

3. 常用的图形函数

图形编程应用广泛，图形界面是当前用户界面的主流，因此呈现给用户的界面就要经过图形编程的包装，以达到更好的效果。常用图形函数有：

Draw_Pixel：画点函数；
Draw_Line：画线函数；
Draw_Circle：画圆函数；
Draw_Rect：绘制矩形；
Draw_Ellipse：绘制椭圆；
Draw_Hline：绘制水平直线；
Draw_Vline：绘制垂直直线；
Draw_Round：绘制圆角矩形。

例如：

```
Draw_Line(screen,400,180,240,300,SDL_MapRGB(screen->format,255,0,0));
for(i=0;i<=640;i+=2)
{
  y=240-120*sin(3.14*i/180);
  Draw_Pixel(screen,i,y,SDL_MapRGB(screen->format,0,255,0));
}
SDL_UpdateRect(screen, 0, 0, 0, 0);                     /*更新整个屏幕*/
```

4. 常用三维绘图函数

glViewport：视觉角度；
glClearColor：清除屏幕时所用的颜色；
glClearDepth：深度缓存；

glDepthFunc：为深度测试设置比较函数；
glShadeModel：阴影模式；
glMatrixMode：选择投影矩阵；
glLoadIdentity：重置当前的模型观察矩阵；
gluPerspective：建立透视投影矩阵；
glTranslatef：指定具体的 x，y，z 值来多元化当前矩阵；
glRotatef：让对象按照某个轴旋转；
glBegin：绘制某个图形；
glColor3f：设置颜色。

例如：

```
void InitGL(int Width, int Height)                        /*初始化 GL 界面*/
{
glViewport(0, 0, Width, Height);                //视觉角度
glClearColor(0.0f, 0.0f, 0.0f, 0.0f);           //背景色设置
glClearDepth(1.0);                              //清除深度缓存
glDepthFunc(GL_LESS);                           //为深度测试选择不同的比较函数
glEnable(GL_DEPTH_TEST);                        //激活深度测试
glShadeModel(GL_SMOOTH);                        //启用阴影平滑
glMatrixMode(GL_PROJECTION);                    //选择投影矩阵
glLoadIdentity();                               //重置投影矩阵
gluPerspective(45.0f,(GLfloat)Width/(GLfloat)Height,1.0f,100.0f);
    // 计算观察窗口的比例和角度等的设置，重置投影矩阵
glMatrixMode(GL_MODELVIEW);
//指明任何新的变换将会影响 modelview matrix（模型观察矩阵）
}
```

10.2 程序设计实例

实例 1 休息提示小程序。该程序创建一个守护进程，每两小时弹出窗口提示用户已连续使用电脑两小时，提醒用户注意休息。程序要求使用 SDL 库，在图形环境下每隔两小时由守护进程创建一个新的进程。源代码设计为：

```
#include<stdlib.h>
#include<stdio.h>
#include<fcntl.h>
#include<SDL.h>
```

```c
#include<SDL_ttf.h>
void init_daemon(){
  pid_t pid;
  int i;
  if ((pid = fork()) < 0){
    perror("创建子进程失败");
    exit(1);
  }
  if (pid > 0) exit(0);
  setsid();
  chdir("/tmp");
  umask(0);
  for (i = 0; i < 1048576; i++) close(i);
}

void main()
{
  SDL_Surface *screen, *text;
  SDL_Color White = {255, 255, 255, 0};
  TTF_Font *Nfont;
  SDL_Rect drect;
  pid_t pid;
  int stat;
  init_daemon();
  while(1)
    {
    sleep(2*60*60);
    if ((pid = fork()) == 0){
      if (SDL_Init(SDL_INIT_VIDEO) < 0){
        fprintf(stderr, "无法初始化:%s\n", SDL_GetError());
        exit(1);
      }
      screen = SDL_SetVideoMode(640, 480, 16, SDL_SWSURFACE);
      if (screen == NULL){
        fprintf(stderr,"无法设置 640X480X16 位色的视频模式:%s\n", SDL_GetError());
        exit(1);
      }
      atexit(SDL_Quit);
      if (TTF_Init() != 0){
```

```
        fprintf(stderr, "无法初始化字体\n");
        exit(1);
     }
     Nfont = TTF_OpenFont("/usr/share/fonts/msyhbd.ttf", 40);
     TTF_SetFontStyle(Nfont, TTF_STYLE_NORMAL);
     text=TTF_RenderUTF8_Blended(Nfont, "YOU have worked for 2 hours.", White);
     TTF_CloseFont(Nfont);
     TTF_Quit();
     drect.x = 15;
     drect.y = 200;
     drect.w = text -> w;
     drect.h = text -> h;
     SDL_BlitSurface(text, NULL, screen, &drect);
     SDL_UpdateRect(screen, 0, 0, 0, 0);
     SDL_FreeSurface(text);
     sleep(5);
     exit(0);
  }
  else{
     waitpid(pid, &stat, 0);
  }
 }
}
```

实例 2 测试人的颜色反应灵敏度。程序的功能是检测被试者的反应能力以及对颜色的敏感程度，能应用于选拔一些对颜色和反应速度有特殊要求的职业（如飞行员）。当测试完成后，程序自动结束，测试者可以在终端看到自己的测试成绩（包括每组测试的反应时间、总耗时和平均反应时间）。程序代码为：

```
#include <SDL.h>
#include <stdlib.h>
#include <stdio.h>
#include <string.h>
#include <time.h>
#include "SDL_draw.h"

#define SCREEN_W 640
#define SCREEN_H 480
#define TEST_NUM 10
#define OUT_TIME 5.0
```

```c
#define DELAY_TIME 500

Uint32 RED,YELLOW,BLUE,GREEN,WHITE,BLACK;
Uint8 *keys;
int color,x,y,rec_w,rec_h,i;
SDL_Surface *screen;
SDL_Event event;
double t[TEST_NUM];

void init_graph();              /*初始化图形环境*/
void define_color();            /*定义颜色*/
void init();                    /*此函数内调用函数init_graph,define_color*/
void rand_data();               /*产生随机数函数*/
void draw_rec();                /*画矩形*/
void output(double t[TEST_NUM]);
int main()
{
init();
for(i=0;i<TEST_NUM;i++)
{
int flag=0;
struct timeval tv1,tv2;
draw_rec();
gettimeofday(&tv1,NULL);
while(flag==0)
  while ( SDL_PollEvent(&event) )
  {
        keys = SDL_GetKeyState(NULL);
        if(keys[SDLK_ESCAPE])
            exit(0);
        else if(keys[SDLK_SPACE])
            goto next;
        else
            switch(color)
            {
             case 1: if(keys['r'])
                    {gettimeofday(&tv2,NULL);
                     flag=1;
                    }
```

```
                            break;
                case 2:  if(keys['y'])
                            {gettimeofday(&tv2,NULL);
                            flag=1;
                             }
                            break;
                case 3:  if(keys['b'])
                            {gettimeofday(&tv2,NULL);
                                    flag=1;
                                    }
                                    break;
                        case 4:  if(keys['g'])
                                    {gettimeofday(&tv2,NULL);
                                    flag=1;
                                    }
                                    break;
                case 5:  if(keys['w'])
                            {gettimeofday(&tv2,NULL);
                            flag=1;
                             }
                            break;
                default: break;
            }
        }
    SDL_Delay(DELAY_TIME);
    t[i]=(tv2.tv_sec-tv1.tv_sec)+(tv2.tv_usec-tv1.tv_usec)/1000000.0;
        next : ;
 Draw_FillRect(screen,x,y,rec_w,rec_h,BLACK);
 }
output(t);
}
void init_graph()
{
 if(SDL_Init(SDL_INIT_VIDEO)<0)
 {    fprintf(stderr, "无法初始化SDL: %s\n", SDL_GetError());
        exit(1);
 }
 screen = SDL_SetVideoMode(SCREEN_W, SCREEN_H, 16, SDL_SWSURFACE);
 if ( screen == NULL )
```

```c
    {
      fprintf(stderr,"无法设置640x480x16位色的视频模式: %s\n", SDL_GetError());
      exit(1);
    }
    atexit(SDL_Quit);
}

void define_color()
{
   RED=SDL_MapRGB(screen->format, 255,0,0);
   YELLOW=SDL_MapRGB(screen->format, 255,255,0);
   BLUE=SDL_MapRGB(screen->format, 0,0,255);
   GREEN=SDL_MapRGB(screen->format, 0,255,0);
   WHITE=SDL_MapRGB(screen->format, 255,255,255);
   BLACK=SDL_MapRGB(screen->format,0,0,0);
}

void init()
{
   init_graph();
   define_color();
   srand((int)time(0));
   for(i=0;i<TEST_NUM;i++)
     t[i]=OUT_TIME;
}

void rand_data()
{
   color=1+(int)(5.0*rand()/(RAND_MAX+1.0));
   x=(int)(540.0*rand()/(RAND_MAX+1.0));
   y=(int)(400.0*rand()/(RAND_MAX+1.0));
   rec_w=30+(int)(80.0*rand()/(RAND_MAX+1.0));
   rec_h=30+(int)(80.0*rand()/(RAND_MAX+1.0));
}

void draw_rec()
{
   rand_data();
   switch(color){
```

```c
    case 1:Draw_FillRect(screen,x,y,rec_w,rec_h,RED);
        break;
    case 2:Draw_FillRect(screen,x,y,rec_w,rec_h,YELLOW);
        break;
    case 3:Draw_FillRect(screen,x,y,rec_w,rec_h,BLUE);
        break;
    case 4:Draw_FillRect(screen,x,y,rec_w,rec_h,GREEN);
        break;
    case 5:Draw_FillRect(screen,x,y,rec_w,rec_h,WHITE);
        break;
  }
  SDL_UpdateRect(screen,0,0,0,0);
}

void output(double t[TEST_NUM])
{
  double sum=0;
  printf("\n*******************************************\n");
  printf("\nYour responce times for each single test are :\n");
  for(i=0;i<TEST_NUM;i++)
  {
    printf("%lf ",t[i]);
    sum=sum+t[i];
  }
  printf("\nThe total time you take is: %lf seconds",sum);
  printf("\nThe average time you take is: %lf seconds\n",sum/TEST_NUM);
  if(sum/TEST_NUM<1.0)
    printf("You are great! Congratulations!!!");
  else
    printf("Never lose heart, try again!!!");
  printf("\n*******************************************\n");
}
```

实例3 程序使用类似于扫雷程序，在屏幕上随机产生"*"号，然后根据产生时间的节点差消失"*"号。

```c
#include<stdio.h>
#include<stdlib.h>
#include<math.h>
struct Pos
```

```c
{
    int x;
    int y;
};

main()
{
    int i = 0,j = 0,k = 0;
    int rand_t = 0,rand_x = 0,rand_y = 0;
    int n = 0;
    int color = 0;
    struct Pos *cursor = NULL;
    int *time = NULL;
    system("clear");
    printf("\033[49;30minput number of points : \033[?25h");
                                //显示光标在屏幕上输出提示
    scanf("%d",&n);
    if(n <= 0)
    return;     //n<0 就直接结束程序
    cursor = (struct Pos*)malloc(n * sizeof(struct Pos));
    time = (int *)malloc(n * sizeof(int));

    for(i = 0;i < n;i++)
    {
        while(1)
        {
            rand_x = abs(rand()) % n + 1;
            rand_y = abs(rand()) % n + 1;
            //产生两个随机数，准备分别作为点 i 的 x,y 坐标.
            for(k = 0;k < i;k++)
            {
                if(cursor[k].x == rand_x && cursor[k].y == rand_y)
                break;
            }
            //检验产生的一对坐标与之前的点是否有重。若有，跳出循环。此时 k<i.
            if(k == i)   //若 k=i,说明新坐标没有与已有的任何点重合。故可以赋值给新的点.
            break;
        }
        while(1)
```

```c
{
rand_t = abs(rand()) % (2 * n) + 1;
for(k = 0;k < i;k++)
{
if(time[k] == rand_t)
break;
}
if(k == i)
break;
}
//同上，令时间节点不同。
j = i;
while(j - 1 >= 0 && time[j - 1] > rand_t)
{
time[j] = time[j - 1];
cursor[j] = cursor[j - 1];
j--;
}
//将各个时间节点排序
time[j] = rand_t;
cursor[j].x = rand_x;
cursor[j].y = rand_y;
}
system("clear");
printf("\033[?25l");

for(i = 0;i < n;i++)
{
color = 30 + abs(rand()) % 9;    //随机产生颜色
printf("\033[%dm\033[%d;%dH*",color,cursor[i].x,cursor[i].y);
}
//产生点

setbuf(stdout,NULL);
sleep(time[0]);
printf("\033[%d;%dH ",cursor[0].x,cursor[0].y);
setbuf(stdout,NULL);

for(i = 1;i < n;i++)
```

```
    {
        sleep(time[i] - time[i - 1]);  //每个点滞留时间为相邻时间节点的差
        printf("\033[%d;%dH ",cursor[i].x,cursor[i].y);
        setbuf(stdout,NULL);
    }
    //消除点

    sleep(4);
    system("clear");
    free(cursor);
    free(time);
    printf("\033[49;30m\033[?25h");
}
```

第 11 章
设备驱动程序设计基础

11.1 知识要点

11.1.1 设备驱动程序概况

1. Linux 环境下设备分为三类

字符设备文件、块设备文件和网络设备文件。

2. 设备文件的查看

使用命令：ls　/dev　–al |more

3. 主设备号与次设备号

主设备号标识设备对应的驱动程序，次设备号由内核使用，用于指向设备。

4. 与设备驱动程序相关的主要数据结构

（1）字符设备驱动程序主要数据结构体：file_operation;
（2）块设备驱动程序主要结数据构体：block_device_operation;
（3）网络设备驱动程序主要数据结构：net_device 和 sk_buff。

11.1.2 字符设备驱动程序

1.模块化字符设备驱动程序的调用流程

模块化加载的驱动程序，用执行 insmod 命令加载驱动模块时，调用函数 module_init("设备初始化函数")注册设备；当执行命令 rmmod 进行卸载设备时，调用函数 module_exit("设备卸载函数")释放设备，如图 11.1 所示。

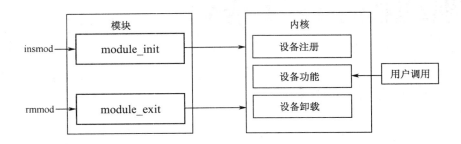

图 11.1 驱动设计的基本流程

2.模块化字符驱动程序设计步骤

（1）定义设备驱动程序 file_operation 结构体相关映射；

例如：
```
struct file_operation gobalvar_fops =
{
 read: gobalvar_read,
 write: gobalvar_write,
};
```

驱动程序在读与写时，会映射到自定义函数 gobalvar_read、gobalvar_write 对设备的读写。

（2）定义设备初始化函数

module_init("设备初始化函数");

在设备初始化函数中，调用 register_chrdev 函数向系统注册设备。

例如：
```
static int __init globalvar_init(void)
{
  if (register_chrdev(MAJOR_NUM, " globalvar ", &globalvar_fops))
  {
    //…注册失败
  }
```

```
    else
    {
      //…注册成功
    }
}
```

（3）定义设备卸载函数

`module_exit("设备卸载函数");`

在设备卸载函数中调用 unregister_chrdev 函数释放设备。

例如：

```
static void __exit globalvar_exit(void)
{
  if (unregister_chrdev(MAJOR_NUM, " globalvar "))
  {
    //…卸载失败
  }
  else
  {
    //…卸载成功
  }
}
```

（4）编译设备

编译成模块时，在 gcc 的命令行里加上这样的参数：-D_KERNEL_ -DMODULE -C

（5）加载模块

当设备驱动程序以模块形式加载时，模块在调用 insmod 命令时被加载，此时的入口地址是 module_init 函数，在该函数中完成设备的注册。

（6）设备读写操作

接着根据用户的实际需要，对相应设备进行读、写等操作。

从内核空间拷贝一块数据到用户空间需要借助函数：

`copy_to_user(void __user *to, const void *from, unsigned long n);`

To 是用户空间的地址；其中，From 是内核空间的地址。

例如：

`copy_to_user(buf, &global_var, sizeof(int));`

`/*buf 用户空间,global_var 内核空间*/`

从用户空间拷贝一块数据到内核空间需要借助函数：

`copy_from_user(void __user *from, const void *to, unsigned long n);`

To 是用户空间的地址；其中，From 是内核空间的地址。

例如：

`copy_from_user(&global_var, buf, sizeof(int));`

`/* global_var 内核空间,buf 用户空间*/`

（7）模块加载

模块用 insmod 命令加载，加载模块时调用函数 module_init；

（8）查看设备

用 lsmod 命令来查看所有已加载的模块状态；

（9）申请设备节点

调用 register_chrdev 成功向系统注册了设备驱动程序后，就可以用 mknod 命令来把设备映射为一个特别文件，其他程序使用这个设备的时候，只要对此特别文件进行操作就行了。

例如：

创建字符设备（c）文件 hello 主设备号为 200、次设备号为 0。

mknod　/dev/hello c　200　0

（10）卸载设备

执行命令 rmmod 时调用函数 module_exit，完成设备的卸载。

11.2　程序设计实例

实例 1　简单的字符设备驱动程序。

1.驱动程序

```
#include <linux/module.h>
#include <linux/init.h>
#include <linux/fs.h>
#include <asm/uaccess.h>
#include <linux/wait.h>
#include <asm/semaphore.h>
MODULE_LICENSE("GPL");

#define MAJOR_NUM 254

static ssize_t globalvar_read(struct file *, char *, size_t, loff_t*);
static ssize_t globalvar_write(struct file *, const char *, size_t, loff_t*);

struct file_operations globalvar_fops =      /*定义字符驱动程序结构体*/
{
  read: globalvar_read,
  write: globalvar_write,
};
```

```c
static int global_var = 0;
static struct semaphore sem;
static wait_queue_head_t outq;
static int flag = 0;

static int __init globalvar_init(void)
{
  int ret;
  ret = register_chrdev(MAJOR_NUM, "globalvar", &globalvar_fops);
  if (ret)
  {
    printk("globalvar register failure");
  }
  else
  {
    printk("globalvar register success");
    init_MUTEX(&sem);
    init_waitqueue_head(&outq);
  }
  return ret;
}

static void __exit globalvar_exit(void)
{
  int ret;
  ret = unregister_chrdev(MAJOR_NUM, "globalvar");
  if (ret)
  {
    printk("globalvar unregister failure");
  }
  else
  {
    printk("globalvar unregister success");
  }
}

static ssize_t globalvar_read(struct file *filp, char *buf, size_t len, loff_t *off)
{
```

```c
    if (wait_event_interruptible(outq, flag != 0))
    {
      return - ERESTARTSYS;
    }

    if (down_interruptible(&sem))
    {
      return - ERESTARTSYS;
    }

    flag = 0;
    if (copy_to_user(buf, &global_var, sizeof(int)))
    {
      up(&sem);
      return - EFAULT;
    }
    up(&sem);
    return sizeof(int);
}

static ssize_t globalvar_write(struct file *filp,const char *buf,size_t len,loff_t *off)
{
    if (down_interruptible(&sem))
    {
      return - ERESTARTSYS;
    }
    if (copy_from_user(&global_var, buf, sizeof(int)))
    {
      up(&sem);
      return - EFAULT;
    }
    up(&sem);
    flag = 1;
    wake_up_interruptible(&outq);
    return sizeof(int);
}

module_init(globalvar_init);
module_exit(globalvar_exit);
```

读端程序用于阻塞地读/dev/globalvar，写端程序用于写/dev/globalvar。只有当后者输入后前者才有返回，否则被阻塞。

2. 读端程序

```c
#include <sys/types.h>
#include <sys/stat.h>
#include <stdio.h>
#include <fcntl.h>
main()
{
  int fd, num;
  fd = open("/dev/globalvar", O_RDWR, S_IRUSR | S_IWUSR);
  if (fd != -1)
  {
    while (1)
    {
      read(fd, &num, sizeof(int));
      printf("The globalvar is %d\n", num);
      if (num == 0)
      {
        close(fd);
        break;
      }
    }
  }
  else
  {
    printf("device open failure\n");
  }
}
```

3. 写端程序

```c
#include <sys/types.h>
#include <sys/stat.h>
#include <stdio.h>
#include <fcntl.h>
main()
{
  int fd, num;
  fd = open("/dev/globalvar", O_RDWR, S_IRUSR | S_IWUSR);
```

```c
  if (fd != - 1)
  {
    while (1)
    {
      printf("Please input the globalvar:\n");
      scanf("%d", &num);
      write(fd, &num, sizeof(int));
      if (num == 0)
      {
        close(fd);
        break;
      }
    }
  }
  else
  {
    printf("device open failure\n");
  }
}
```

实例 2 Mini2410 下基础驱动开发，LED 控制。

在驱动中添加 write 形式方法，修改应用程序，输入一个 4 位的 1 或者 0，调用 write 方法，使得为 1 的位亮灯，为 0 的位灭灯，例如，输入 1010，实验箱上的四盏灯则"亮灭亮灭"。

```c
#include <linux/config.h>
#include <linux/module.h>
#include <linux/kernel.h>
#include <linux/fs.h>
#include <linux/init.h>
#include <linux/devfs_fs_kernel.h>
#include <linux/miscdevice.h>
#include <linux/delay.h>
#include <asm/irq.h>
#include <asm/io.h>
#include <asm/arch/regs-gpio.h>
#include <asm/hardware.h>
#include <asm/uaccess.h>
#define DEVICE_NAME"led"
#define LED_MAJOR 233
#define LED_BASE(0xE1180000)
```

```c
unsigned char status = 0xff;
static int eduk4_led_ioctl(struct inode *inode, struct file *file, unsigned int cmd, unsigned long arg)
{
// unsigned char status = 0xff;
switch(cmd) {
case 0:
case 1:
if (arg > 4) {
return -EINVAL;
}
// status = inb(LED_BASE);
if(0 == cmd){
status &= ~(0x1 << arg);
}else if(1 == cmd){
status |= (0x1 << arg);
}
outb(status, LED_BASE);
return 0;
default:
return -EINVAL;
}
}
static ssize_t eduk4_led_read(struct file *file, char __user *buf, size_t count, loff_t *ppos){
printk("\nE201102032 LiDandan \n3090101023 JiangYing\n");
}
static ssize_t eduk4_led_write(struct file *file, char __user *buf, size_t count, loff_t *ppos){
unsigned char status = 0xff;
printk("into\n");
int ret,i;
char from_user_data[4] ;
ret = copy_from_user(from_user_data, buf, sizeof(from_user_data));
status = inb(LED_BASE);
for(i=0;i<4;i++)
{
if(from_user_data[i]=='0')
status &= ~(0x1 << i);
```

```c
        else if(from_user_data[i]=='1')
        status |=(0x1 << i);
    }
    outb(status, LED_BASE);
    return 0;
}
static struct file_operations eduk4_led_fops = {
.owner=THIS_MODULE,
.ioctl=eduk4_led_ioctl,
.read =    eduk4_led_read,
.write=    eduk4_led_write,
};
static int __init eduk4_led_init(void)
{
int ret;
unsigned char status;
ret = register_chrdev(LED_MAJOR, DEVICE_NAME, &eduk4_led_fops);
if (ret < 0) {
printk(DEVICE_NAME " can't register major number\n");
return ret;
}
    devfs_mk_cdev(MKDEV(LED_MAJOR, 0), S_IFCHR | S_IRUSR | S_IWUSR | S_IRGRP, DEVICE_NAME);
    status = inb(LED_BASE);
    outb(status | 0xff,LED_BASE);
    printk(DEVICE_NAME " initialized\n");
    return 0;
}
static void __exit eduk4_led_exit(void)
{
unsigned char status;
    status = inb(LED_BASE);
outb(status | 0xff,LED_BASE);
printk(DEVICE_NAME " remove\n");
devfs_remove(DEVICE_NAME);
unregister_chrdev(LED_MAJOR, DEVICE_NAME);
}
module_init(eduk4_led_init);
module_exit(eduk4_led_exit);
```

MODULE_LICENSE("BSD/GPL");

led_test

```c
#include <stdio.h>
#include <stdlib.h>
#include <unistd.h>
#include <sys/ioctl.h>
#include <sys/types.h>
#include <sys/stat.h>
#include <fcntl.h>
#include <sys/select.h>
#include <sys/time.h>
static int led_fd;
int ccc,kk,mmm;
int ret[4];
int led[4];
int main(int argc, char **argv)
{
char t[4];
int i=0;
// open device
led_fd = open("/dev/led", O_RDWR);
if (led_fd < 0) {
perror("open device led");
exit(1);
}
printf("Please look at the leds\n");
read(led_fd,ret,kk);
scanf("%s",t);
printf("%d\n",write(led_fd, t, 4));
close(led_fd);
return 0;
}
```

实例3 阅读并理解下列程序。

（1）驱动程序文件名 driver.c

```c
#include <linux/module.h>
#include <linux/kernel.h>
#include <linux/init.h>
#include <linux/fs.h>
#include <asm/uaccess.h>
```

```c
MODULE_LICENSE("GPL");
#define ID  (*(volatile unsigned * )0x3ff5008)      //GPIO数据寄存器
#define IM  (*(volatile unsigned * )0x3ff5000)      //GPIO模式寄存器
static int moto_write(struct file*,char*,int,loff_t*);
static int major = 212;                 //定义设备号为212
static char moto_name[] = "moto1";      //定义设备文件名为moto

static void delay_moto(unsigned long counter)      //延时函数
{
    unsigned int a;
    while(counter--)
     {
        a = 400;
        while(a--)
          ;
     }
}


static struct file_operations moto1_fops=   //声明file_operations结构
{
write : (void(*)) moto_write,
};

//驱动程序初始化函数
static int __init moto_init_module(void)
{
int retv;
retv = register_chrdev(major, moto_name, &moto1_fops);
if(retv<0)
{
    printk("<1>Register fail! \n");
    return retv;
}
if (major==0)
    major = retv;
printk("moto1 regist success\n");
return 0;
}
//驱动程序退出函数
```

```c
static void motodrv_cleanup(void)
{
int retv;
retv = unregister_chrdev(major, moto_name);
if(retv<0)
{
printk("<1>Unregister fail! \n");
return;
}
printk("<1>MOTODRV: Good-bye! \n");
}

static int moto_write(struct file*moto_file,char*buf,int len, loff_t* loff)
{
unsigned long on;
IM = 0xff;
if(copy_from_user((char * )&on, buf, len))
    return -EFAULT;
if(on)//快转
    {
    ID |= 0x10;          //set A_bit
    delay_moto(20);
    ID &= 0xffffffef;    //clear A_bit
    ID |= 0x20;          //set B_bit
    delay_moto(20);
    ID &= 0xffffffdf;    //clear B_bit
    ID |= 0x40;          //set C_bit
    delay_moto(20);
    ID &= 0xffffffbf;    //clear C_bit
    ID |= 0x80;          //set D_bit
    delay_moto(20);
    ID &= 0xffffff7f;    //clear D_bit
    }
else
    {
    ID |= 0x10;          //set A_bit
    delay_moto(40);
    ID &= 0xffffffef;    //clear A_bit
    ID |= 0x20;          //set B_bit
```

```
        delay_moto(40);
        ID &= 0xffffffdf;   //clear B_bit
        ID |= 0x40;         //set C_bit
        delay_moto(40);
        ID &= 0xffffffbf;   //clear C_bit
        ID |= 0x80;         //set D_bit
        delay_moto(40);
        ID &= 0xffffff7f;   //clear D_bit
    }
    return len;
}
module_init(moto_init_module);
module_exit(motodrv_cleanup);
```

（2）主程序文件名 app.c

```
#include <fcntl.h>
void delay();
int main(void)
{
int fd;
unsigned int on,i,j;//on=0 代表慢，on=1 代表快
fd = open("/dev/moto1", O_RDWR);     //打开设备 moto1
while(1)
{
    i = (*(volatile unsigned *)0x03FF5008);//判断连接到 GPIO 的按钮是否按下
    i &= 0x00003000;
    j = i;
    i &= 0x00001000;
    j &= 0x00002000;
    if(i)         //  启停键按下
    {
            if(j)          //快慢键按下
        {
        on = 0x01;
        write(fd, (char *)&on, 4);
        }
        else
        {
        on = 0x0;
        write(fd, (char *)&on, 4);
```

```
            }
        }
    }
    close(fd);
    return 0;
}
void delay()              //延时函数
{
    long int i=1000000;
    while(i--);
}
```

第 12 章 串行通信

12.1 知识要点

12.1.1 串行通信

1. 串口设备名

在 Linux 中所有设备文件一般都位于/dev 目录下,串口一与串口二对应的设备为/dev/ttyS0 及/dev/ttyS1。

2. 串行通信分类

串行通信可分为同步通信与异步通信。

同步通信是一种连续串行传送数据的通信方式,一次通信只传送一帧信息。此信息帧由同步字符、数据字符和校验字符(CRC)组成。同步通信的缺点是要求发送时钟和接收时钟保持严格的同步。

异步通信在发送字符时,所发送的字符之间的间隙可以是任意的。在异步通行中有两个比较重要的指标:字符帧格式和波特率。字符帧由发送端逐帧发送,通过传输线被接收设备逐帧接收。发送端和接收端可以由各自的时钟来控制数据的发送和接收,这两个时钟源彼此独立,互不同步。接收端检测到起始位时,确定发送端已开始发送数据。每当接收端收到字符帧中的停止位时,就知道一帧字符已经发送完毕。

3. 终端工作模式

（1）规范模式：在 c_lflag 中设置 ICANNON 标志，基于行处理，用户在输入行结束符时，系统调用 read 才能读到用户的输入，并且调用一次 read 最多只能读取一行；

（2）非规范模式：在 c_lflag 中清除 ICANNON 标志，所有的输入是即时生效的；

12.1.2 串行通信程序设计流程

（1）打开通信端口，保存原端口的设置；
（2）配置端口波特率、字符大小等参数，设置串口参数；
（3）对该串行口进行数据读写操作；
（4）通信结束后，关闭串口。

12.1.3 串行通信程序设计步骤

1. 接收端

（1）打开 PC 的 COM1 端口

如果以读写的方式打开 COM1 端口，语句可写为：
```
fd=open("/dev/ttyS0",O_RDWR | O_NOCTTY);
```
（2）取得当前串口值，并保存至结构体变量 oldtio
```
tcgetattr(fd,&oldtio);
```
（3）串口结构体变量 newtio 清零
```
bzero(&newtio,sizeof(newtio));
```
（4）设置串口参数

1）假定设置波特率为 BAUDRATE，8 个数据位，忽略任何调制解调器状态，同时启动接受器，CLOCAL、CREAD 分别用于本地连接和接收使能。
```
newtio.c_cflag=BAUDRATE |CS8 |CLOCAL|CREAD;
```
2）忽略奇偶校验错误。
```
newtio.c_iflag=IGNPAR;
```
3）设输出模式非标准型，同时不回应。
```
newtio.c_oflag=0;
```
4）启用正规模式。
```
newtio.c_lflag=ICANON;
```
（5）刷新在串口中的输入输出数据
```
tcflush(fd,TCIFLUSH);
```
参数 TCIFLUSH 表示刷新收到的数据，但是不读取

（6）设置当前的串口参数为 newtio

```
tcsetattr(fd,TCSANOW,&newtio);
```
参数 TCSANOW 表示改变端口配置并立即生效

（7）读取缓存中的数据
```
read(fd,buf,255);
```
表示从端口读取 255 个字符存放到 buf 中

（8）关闭串口
```
close(fd);
```

（9）恢复旧的端口参数
```
tcsetattr(fd,TCSANOW,&oldtio);
```

2. 发送端

（1）打开 PC 的 COM2 端口
```
fd=open("/dev/ttyS1",O_RDWR | O_NOCTTY);
```

（2）取得当前串口值，并保存至 oldtio
```
tcgetattr(fd,&oldtio);
```

（3）串口结构体变量 newtio 清零
```
bzero(&newtio,sizeof(newtio));
```

（4）设置串口参数

1）设置波特率为 BAUDRATE，8 个数据位，忽略任何调制解调器状态，同时启动接受器。
```
newtio.c_cflag=BAUDRATE |CS8 |CLOCAL|CREAD;
```

2）忽略奇偶校验错误。
```
newtio.c_iflag=IGNPAR;
```

3）设输出模式非标准型，同时不回应。
```
newtio.c_oflag=0;
```

4）启用正规模式。
```
newtio.c_lflag=ICANON;
```

（5）刷新在串口中的输入输出数据
```
tcflush(fd,TCIFLUSH);
```

（6）设置当前的串口为 newtio
```
tcsetattr(fd,TCSANOW,&newtio);
```
参数 TCSANOW 表示改变端口配置并立即生效

（7）向串口写入数据，储存在缓存中
```
write(fd,s1,1);
```
表示向串口写入长度为 1 的数据 s1

（8）关闭串口
```
close(fd);
```

（9）恢复旧的端口参数
```
tcsetattr(fd,TCSANOW,&oldtio);/*恢复旧的端口参数*/
```

12.2 程序设计实例

实例1 串口设置实例，一个简单的打字测速程序。程序读入一个文本文件，并在屏幕上显示出该文件的内容，然后读入用户的输入。程序可以计算正确率以及所花费的总时间。该程序由于测试的是按键的反应时，在输入时不能用回车，因而需要通过输入、输出端口的属性设置，输入字符被收到后就直接传送及显示，此时需要取消ICANON、添加ECHO属性，即newt.c_lflag &=(~ICANON) & ECHO；程序代码：

```
#include <stdio.h>
#include <sys/time.h>
#include <unistd.h>
#include <termios.h>

int getch()
{
struct termios oldt, newt;
int ch;
tcgetattr(STDIN_FILENO, &oldt);
newt = oldt;
newt.c_lflag &= (~ICANON) & ECHO;
tcsetattr(STDIN_FILENO, TCSANOW, &newt);
ch = getchar();
tcsetattr(STDIN_FILENO, TCSANOW, &oldt);
return ch;
}

int main(int argc, char *argv[])
{
FILE *fp;
int i = 0, count = 0, correct = 0;
int text[1000], ch;
struct timeval tv1, tv2;
struct timezone tz;
if (argc == 1)
{
    printf("Usage: simple3 <Filename>\n");
```

```c
        return 0;
    }
    if (!(fp = fopen(argv[argc - 1], "r")))
    {
        printf("Cannot open file!\n");
        return 0;
    }
    while (1)
    {
        ch = fgetc(fp);
        if (feof(fp)) break;
        text[count++] = ch;
        putchar(ch);
    }
    fclose(fp);
    gettimeofday(&tv1, &tz);
    for (i = 0; i < count; i++)
    {
        ch = getch();
        if (ch == text[i]) correct++;
    }
    gettimeofday(&tv2, &tz);
    printf("Correct characters: %d/%d\n", correct, count);
    printf("Time usage: %ds\n", tv2.tv_sec - tv1.tv_sec);
    return 0;
}
```

实例2 键盘监听程序。程序通过对键盘的终端属性放置，用以实现的键盘监听功能。

程序代码：

```c
#include<stdio.h>
#include<termios.h>
#include<term.h>
#include<curses.h>
#include<unistd.h>
static struct termios initial_settings, new_settings;
static int peek_character = -1;
//function prototype
void init_keyboard();
void close_keyboard();
int kbhit();
```

```c
int readch();
//main
int main()
{
int ch =0;
init_keyboard();
while(ch != 'q'){
printf("looping\n");
sleep(1);
if(kbhit())
{
ch = readch();
printf("You hit %c\n",ch);
}
}
close_keyboard();
return 0;
}
void init_keyboard()
{
tcgetattr(0, &initial_settings);
new_settings = initial_settings;
new_settings.c_lflag &= ~ICANON;
new_settings.c_lflag &= ~ECHO;
new_settings.c_lflag &= ~ISIG;
new_settings.c_cc[VMIN] = 1;
new_settings.c_cc[VTIME] = 0;
tcsetattr(0, TCSANOW, &new_settings);
}
void close_keyboard()
{
tcsetattr(0, TCSANOW, &initial_settings);
}
int kbhit()
{
char ch;
int nread;
if(peek_character!=-1)
return 1;
```

```c
new_settings.c_cc[VMIN]=0;
tcsetattr(0, TCSANOW, &new_settings);
nread = read(0,&ch, 1);
new_settings.c_cc[VMIN] = 1;
tcsetattr(0, TCSANOW, &new_settings);
if(nread ==1)
{
peek_character = ch;
return 1;
}
return 0;
}
int readch()
{
char ch;
if(peek_character != -1)
{
ch = peek_character;
peek_character =-1;
return ch;
}
read(0, &ch, 1);
return ch;
}
```

实例3 通过计算机的 COM1 和 COM2 进行通信,利用 RS-232 来传送,COM1 为传送端,COM2 为接收端,要求运用串行通信完成两个串行口之间对计算机任意一文件的传送,发送端打开文件并读取文件内容,然后将文件的内容传送到接收端,接收端将文件内容打印到屏幕上。接收文件结束时输入的终止字符为'#',其通信格式为:9600, n, 8, 1。源程序如下:

发送端程序 12-3-s.c:

```c
#include <stdio.h>
#include<sys/types.h>
#include<sys/stat.h>
#include<fcntl.h>
#include<termios.h>
#define BAUDRATE B9600
#define MODEMDEVICE "/dev/ttyS0"
#define STOP '#'
int main()
```

```c
{
    int fd,c=0,res,fdsrc,nbytes;
    struct termios oldtio,newtio;
    char ch,s1[256],car[20];
    printf("start...\n");
    fd=open(MODEMDEVICE,O_RDWR | O_NOCTTY);
    if(fd<0)
    {
        perror(MODEMDEVICE);
        exit(1);
    }
    printf("open...\n");
    tcgetattr(fd,&oldtio);
    bzero(&newtio,sizeof(newtio));
    newtio.c_cflag=BAUDRATE|CS8|CLOCAL|CREAD;
    newtio.c_iflag=IGNPAR;
    newtio.c_oflag=0;
    newtio.c_lflag=ICANON;
    tcflush(fd,TCIFLUSH);
    tcsetattr(fd,TCSANOW,&newtio);
    printf("writing...\n");
    printf("请输入需要传送的文件名：");
    while(1)
    {
            scanf("%s",car);
            ch=getchar();
            fdsrc=open(car,O_RDONLY);
            if(fdsrc<0)
                exit(1);
            while((nbytes=read(fdsrc,s1,255))>0)
                res==write(fd,s1,nbytes);
            ch=getchar();
            s1[0]=ch;
            s1[1]='\n';
            res=write(fd,s1,2);
            break;
    }
}
    printf("close...\n");
```

```
        close(fdsrc);
    close(fd);
    tcsetattr(fd,TCSANOW,&oldtio);
    return 0;
}
```

接收端程序 12-3-r.c

```c
#include <stdio.h>
#include <sys/types.h>
#include<fcntl.h>
#include<termios.h>
#define BAUDRATE B9600
#define MODEMDEVICE "/dev/ttyS1"
int main()
{
int fd,c=0,res;
struct termios oldtio, newtio;
char buf[256];
printf("start ...\n");
fd=open(MODEMDEVICE,O_RDWR | O_NOCTTY);
if(fd<0)
{
    perror(MODEMDEVICE);
    exit(1);
}
printf("open...\n");
tcgetattr(fd,&oldtio);
bzero(&newtio,sizeof(newtio));
newtio.c_cflag=BAUDRATE |CS8 |CLOCAL|CREAD;
newtio.c_iflag=IGNPAR;
newtio.c_oflag=0;
newtio.c_lflag=ICANON;
tcflush(fd,TCIFLUSH);
tcsetattr(fd,TCSANOW,&newtio);
printf("reading...\n");
while(1)
{
    res=read(fd,buf,255);
    buf[res]=0;
    printf("res=%d vuf=%s\n",res,buf);
```

```
        if(buf[0]=='#') break;
}
printf("close...\n");
close(fd);
tcsetattr(fd,TCSANOW,&oldtio);
return 0;
}
```

实践部分

Linux程序设计实践与编程技巧

Linux 程序设计实验报告 1

——操作系统基本命令使用

一、实验目的

1.通过对 Emacs、vi、vim、gedit 文本编辑器的使用，掌握在 Linux 环境下文本文件的编辑方法；

2.通过对常用命令 mkdir、cp、cd、ls、mv、chmod、rm 等文件命令的操作，掌握 Linux 操作系统中文件命令的用法。

二、实验任务与要求

1.emacs 的使用，要求能新建、编辑、保存一个文本文件；
2.vi 或 vim 的使用，要求能新建、编辑、保存一个文本文件；
3.gedit 的使用，要求能新建、编辑、保存一个文本文件；
4.掌握 mkdir、cd 命令的操作，要求能建立目录、进入与退出目录；
5.掌握 cp、ls、mv、chmod、rm 命令的操作，要求能拷贝文件、新建文件、查看文件、文件重命名、删除文件等操作。

三、实验工具与准备

PC 机、Linux Redhat Fedora Core6 操作系统。

四、实验步骤与操作指导

任务 1 学习 emacs 的使用，要求能新建、编辑、保存一个文本文件。
（1）启动 emacs；
（2）输入以下 C 程序；
（3）保存文件为 kk.c；
（4）用 emacs 打开文件 kk.c；
（5）修改程序；
（6）另存为文件 aa.txt 并退出。

任务 2 vi 或 vim 的使用，要求能新建、编辑、保存一个文本文件。

（1）点击"应用程序"→"附件"→"终端"，打开终端，在终端输入命令：
[root@localhost root]#vi kk.c
按 i 键，进入插入状态。

（2）输入以下 C 程序：

```
#include<stdio.h>
int main( )
{
 printf("Hello world!\n");
 return 0;
}
```

此时可以用 Backspace、→、←、↑、↓键编辑文本。

（3）保存文件为 kk.c。
按 Esc 键，进入最后行状态，在最后行状态输入：wq 保存文件，退出 vi。

（4）用 vi 打开文件 kk.c，输入命令：
[root@localhost root]#vi kk.c

（5）修改程序为：

```
#include<stdio.h>
int main( )
{
  printf("*****************\n");
  printf("Hello world!\n");
  printf("*****************\n");
  return 0;
}
```

（6）按 Esc 键，进入最后行状态，在最后行状态输入：wq aa.txt 保存文件，如图 1 所示，另存为文件 aa.txt 并退出 vi。

图 1　程序编辑环境

任务 3 gedit 的使用，要求能新建、编辑、保存一个文本文件。
（1）启动 gedit，点击"应用程序"→"附件"→"文本编辑器"，打开文本编辑器，如图 2 所示。

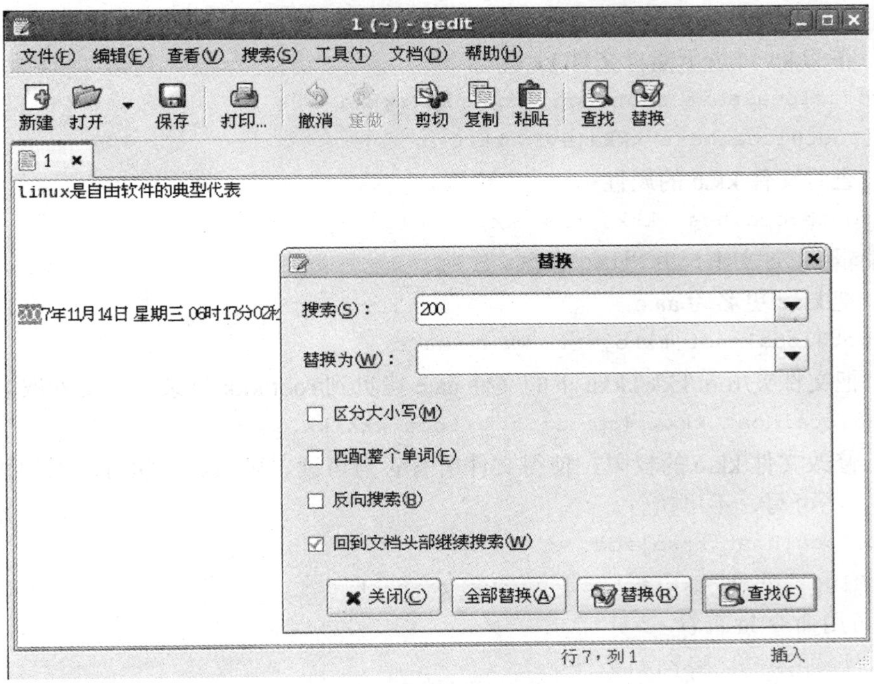

图 2 文本编辑器 gedit

（2）输入以下 C 程序；
（3）保存文件为 kk.c；
（4）用 gedit 打开文件 kk.c；
（5）修改程序；
（6）另存为文件 aa.txt 并退出。

任务 4 掌握 mkdir、cd 命令的操作，要求能建立目录、进入与退出目录。
（1）打开"终端"应用程序"→"附件"→"终端"，在终端用命令新建目录 kkk；
　　[root@localhost root]#mkdir kkk
（2）进入目录 kkk，并在 kkk 目录下新建目录 kkka，进入 kkka 目录；
　　[root@localhost root]#cd kkk
　　[root@localhost kkk]#mkdir kkka
　　[root@localhost kkk]#cd kkka
　　[root@localhost kkka]#
（3）执行命令 cd ..命令，然后再进入 kkka 目录，输入命令 cd、cd /etc，观察其结果。
　　[root@localhost kkka]#cd ..
　　[root@localhost kkk]#cd kkka
　　[root@localhost kkka]#cd
　　[root@localhost root]#cd /etc
　　[root@localhost etc]#

任务 5 掌握 cp、ls、mv、chmod、rm 命令的基本操作，要求能拷贝文件、新建文件、查看文件的权限、修改文件以及删除文件。

（1）在 kkka 目录下建立文件 kk.c；

　[root@localhost root]#cd /root/kkk/kkka

　　[root@localhost kkka]#vi　kk.c

（2）查看文件 kk.c 的属性；

　[root@localhost kkka]#ls　kk.c　-l

　编辑 kk.c 文本，并用:wq 存盘。

（3）把 kk.c 更名为 aa.c；

　[root@localhost kkka]#mv　kk.c　aa.c

（4）把文件夹/root/kkk/kkka 下的文件 aa.c 拷贝到/root/kkk 目录下，文件取名为 kk.c。

[root@localhost kkka]#cp　aa.c　/root/kkk/kk.c

（5）修改文件 kk.c 的权限，使得文件所有者为可读、可写、可执行，对组内人及其他人可读、不可写、不可执行。

[root@localhost kkka]#cd ..

[root@localhost kkk]#chmod　u=rwx,go=r kk.c

此时可用命令 ls 查看

[root@localhost kkk]#ls　-l

（6）删除文件与文件夹。

删除 kkka 文件夹下的文件 aa.c

[root@localhost kkk]#rm kkka/aa.c

查看文件夹 kkka 下否删除了文件 aa.c

[root@localhost kkk]ls kkka/aa.c　-l

删除 kkka 文件夹下

[root@localhost kkk]#rmdir　kkka

查看是否删除了文件夹

[root@localhost kkk]ls kkka　-l

（7）新建一个 linux_d 目录，并设置它的权限为 666。

（8）在指定的目录中搜索文件，利用 find 命令搜索含有通配符的文件*.c。

（9）练习命令:ping\netstat\mount\ifconfig\

（10）在根目录下用 find 查找.c 文件。

（11）练习检查磁盘命令 fdisk 的使用。列出结果中有关/dev/sd 的磁盘信息。

　　　/sbin/fdisk -l |grep /dev/sd

（12）设置当前的时间为 2013 年 10 月 01 日 10 点 23 分。

（13）在后台运行命令 gedit，并用命令 kill 杀死 gedit 的进程。

（14）检查磁盘，列出目录/dev/sd 的磁盘信息。

（15）查找/usr/sbin 及/usr/bin/两个目录中所有的 C 语言程序。

（16）统计当前目录下 txt 文件的总字节数。

（17）把 ps 命令的标准输出结果输入给 sort，经过排序后结果被保存到 pssort.out 中。

178

五、实验结果记录

六、实验结果分析

七、实验心得

Linux 程序设计实验报告 2

——Shell 程序设计 1

一、实验目的

1. Shell 程序设计中变量的使用；
2. 理解通道的概念并初步掌握它的使用方法；
3. 掌握算术操作、字符串操作、逻辑操作、文件操作；
4. 掌握 if then fi、if then elif fi、case、while、for 等控制语句；
5. 在 shell 脚本中使用函数。

二、实验任务与要求

1. 观察变量$#、$0、$1、$2、$3、$@的含义；
2. Shell 程序设计中文件与文件夹的判断；
3. 顺序、分支、循环程序的设计；
4. 菜单程序的编写。

三、实验工具与准备

PC 机、Linux Redhat Fedora Core6 操作系统。

四、实验步骤与操作指导

任务 1 调试下列 shell 程序，写出变量$#、$0、$1、$2、$3、$@的含义。

```
#! /bin/bash
echo "程序名:$0"
echo "所有参数: $@"
echo "前三个参数:$1 $2 $3"
shift
echo "程序名:$0"
echo "所有参数: $@"
```

```
echo "前三个参数:$1 $2 $3"
shift 3
echo "程序名:$0"
echo "所有参数：$@"
echo "前三个参数:$1 $2 $3"
exit 0
```

修改程序，使用变量$#，程序运行时从键盘输入文件名，判断文件是否存在，如果存在，显示文件内容。

提示：

```
read  DORF
if [ -d $DORF ]
then
  ls $DORF
elif [ -f $DORF ]
```

任务2 编写一个 shell 程序，此程序的功能是：显示 root 下的文件信息，然后建立一个 kk 的文件夹，在此文件夹下新建一个文件 aa，修改此文件的权限为可执行。

提示：

1. 进入 root 目录：cd /root
2. 显示 root 目录下的文件信息：ls   l
3. 新建文件夹 kk：mkdir kk
4. 进入 root/kk 目录：cd kk
5. 新建一个文件 aa：vi aa   #编辑完成后需手工保存
6. 修改 aa 文件的权限为可执行：chmod +x aa
7. 回到 root 目录：cd /root

请修改程序，所建立的目录名从键盘输入，把/root 下的所有文件信息保存在 aa 文件中。

任务3 调试下列 shell 程序，此程序的功能是：利用内部变量和位置参数编写一个名为 test2 的简单删除程序，如删除的文件名为 a，则在终端输入的命令为 test a。

提示：
除命令外至少还有一个位置参数，即$#不能为0，删除的文件为$1。

（1）用 vi 编辑程序
[root@localhost bin]#vi test2

```
#!/bin/sh
if test $# -eq 0
then
  echo "Please specify a file!"
else
  gzip $1 //先对文件进行压缩
  mv $1.gz $HOME/dustbin //移动到回收站
  echo "File $1 is deleted !"
fi
```

（2）请修改程序，查看回收站中的文件，从键盘输入回收站中的某一文件，把此文件恢复到/home 目录下。

（3）删除垃圾箱中的所有文件。

任务4 调试下列程序，程序的主要思想是用 while 循环求 1 到 100 的和。

（1）用 gedit 编辑脚本程序 test12

```
[root@localhost bin]#gedit test12
total=0
num=0
while((num<=100));do
    total=`expr $total + $num
   ((num+=1))
done
echo "The result is $total"
```

（2）用 for 语句完成以上求和。

（3）编写 shell 程序计算 1+1/2+1/3+1/4+…+1/n

任务5 调度下列程序，使用 shell 编写一个菜单，分别实现列出以下内容：①目录内容；②切换目录；③创建文件；④编辑文件；⑤删除文件的功能。在此例中将用到循环语句 until、分支语句 case、输入输出语句 read 和 echo。

```
#! /bin/bash
until
    echo "(1)List you selected directory"
    echo "(2)Change to you selected directory"
    echo "(3)Creat a new file"
    echo "(4)Edit you selected file"
    echo "(5)Remove you selected file"
    echo "(6)Exit Menu"
    read input
```

```
            if test $input = 6 then
                exit 0
            fi
    do
        case $input in
            1) ls;;
            2) echo -n "Enter target directory:"
               read dir
               cd $dir
               ;;
            3) echo -n "Enter a file name:"
               read file
               touch $file
               ;;
            4) echo -n "Enter a file name:"
               read file
               vi $file
               ;;
            5) echo -n "Enter a file name:"
               read file
               rm $file
               ;;
            *) echo "Please selected 1\2\3\4\5\6 " ;;
        esac
    done
```

（1）修改以上程序，用菜单形式完成算术四则混合运算。

（2）修改以上程序，用菜单形式完成7种电脑图形游戏。

任务6 调试下列程序。

```
#!/bin/bash
xx=0
func(){
dire=${PWD%/*}
for file in $(ls); do
if [ -f "$file" ]; then
    i=xx
    while [ "$i" -gt "0" ]; do
        echo /c "+"
    done
```

```
        echo /c "-"
        echo ${dire}"/$file"
elif [ -d "$file" ]; then
        cd "$file"
        dire=${PWD %/*}
        xx=$(($xx+1))
        func;
        cd ..
    fi
done
}
func
```

分析程序的运行结果，此程序的功能类似于 Windows 的什么命令。

五、实验结果记录

六、实验结果分析

七、实验心得

Linux 程序设计实验报告 3

——Shell 程序设计 2

一、实验目的

1. 提高 Shell 程序编程的技巧；
2. 提升 Shell 综合编程能力。

二、实验任务与要求

1. 菜单的实现；
2. 遍历所有以某目录为祖先的文件；
3. 分支与循环、随机数在游戏程序中的应用；
4. 定时检查存文件储空间的变化；
5. 消息框程序设计；
6. 菜单界面程序设计。

三、实验工具与准备

PC 机、Linux Redhat Fedora Core6 操作系统。

四、实验步骤与操作指导

任务 1 编写下列程序。程序的功能是实现如下菜单及菜单所表示的功能。
请输入您要清空的文件类型
　　1—文件夹
　　2—其他文件
　　3—所有文件
　　0—没想好，先退出

任务 2 补充完整以下 Shell 程序，已有函数 list 的功能是遍历所有以该目录为祖先的文件，要求输入一个目录名，大致实现命令 "ls –R" 的功能。

```sh
#!/bin/sh
list()
{
   cd $1
   ls -l
   for i in $(ls $1)
   do
    if [ -d $i ]
      then
      directory="$1/$i"
      echo "The directory $directory is a subdirectory of $1, which includes:"
      list $directory
      cd $1
    fi
   done
}
```

任务3 调试下列 Shell 程序,这是一个类似于扑克牌的小游戏。请阅读程序,写出这个游戏的玩法。

```bash
#!/bin/bash
for i in $( seq 1 54 )              #初始化牌
do
        ok[$i]=1
done
max=0                               #初始化赢家
for i in $(seq 1 4 )                #游戏开始
do
        echo "Player $i"
        a=$((RANDOM%53+1))          #玩家 i 的回合
        while [[ ${ok[$a]} -eq 0 ]]; #抽牌
        do
        a=$((RANDOM%53+1))
        done
        ok[$a]=0
        if [ $a -gt $max ]; then
             max=$a
             num=$i
        fi
        HuaSe=$((a%4))              #生成花色
```

```
        case $HuaSe in
                0)HS='C';;          #草花
                1)HS='D';;          #方块
                2)HS='H';;          #红心
                3)HS='S';;          #黑心
        esac
        DaXiao=$((a/4+1))           #生成牌值
        case $DaXiao in
                2|3|4|5|6|7|8|9|10) echo $HS $DaXiao;;
                11)echo $HS 'J';;
                12)echo $HS 'Q';;
                13)echo $HS 'K';;
                1 )echo $HS 'A';;
                14)echo 'small JOKE ';; #小王
                15)echo 'big   JOKE ';; #大王
        esac
        read ll
done
echo "Player $num win !"             #游戏结果
```

任务4 编写下列 Shell 程序。程序每隔 5 分钟检查一下当前用户（假设用户名为 liujh）是否有新的邮件，若有则提示用户。

 提示：

题目的关键问题是获取 5 分钟前后此文件夹的存储容量，如果邮件箱中存储容量发生变化，则可判断有新邮件到达。提示关键语句：

```
count1=`ls -l /var/mail/liujh|awk '{print $5}'`
echo $count1
sleep 300        #隔5分钟检测一次
count2=`ls -l /var/mail/liujh|awk '{print $5}'`
echo $count2
if [ $count1 -eq $count2 ]
```

任务5 调试下列程序，写出程序调试结果并改写程序为菜单控制。
```
#! /bin/sh
dialog --title "Start" --msgbox
dialog --title "Welcome to the Start Menu."
dialog --title "Confirm" --yesno "Do you want to enter the Menu?" 9 18
```

```
    if [ $? != 0]; then
            dialog --infobox "Welcome!"
            sleep 2
            dialog --clear
            exit 0
    fi
    dialog --menu "MENU" 12 24 3 1 "CO Player" 2 "Radio" 3 "Sound Control" 2>_1.txt
    M_O = $(cat _1.txt)
    if [ "$M_O" = "1" ] ; then
       !gnome-cd
    elif [ "$M_O" = "2" ] ; then

       !gnome-sound-recorder
    else
       !gnome-volume-control
    fi
    exit 0
```

任务6 调试下列程序。

程序功能设计一个简单的调查问卷菜单界面，调查学生一些个人信息。此程序运用了 GUI 图形化界面、使用了条件语句，case 语句，使用户可以对菜单进行选择，并将选择结果保存在临时文件中，便于进一步操作计算。

程序的源代码为：

```
#!/bin/sh
    gdialog --title "Questionnaire" --yesno "Will you participate in this survey?" 9 18
    if( $? != 0 ); then
            gdialog --infobox " Thank you "
            sleep 1
            gdialog --clear
            exit 0
    fi
    gdialog --title "Questionnaire" --msgbox "This is a survey about your personal information" 9 18
    gdialog --title "Questionnaire" --inputbox "Enter your name" 9 18 2>name.txt
    _name=$(cat name.txt)
    gdialog --menu "$_name, what is your major?" 9 18 3 1 "Liberal Art" 2 "Science" 3 "Computer Science" 2>choice.txt
    _choice=$(cat choice.txt)
```

```
case "$_choice" in
    1) gdialog --title "Questionnaire" --msgbox "Good choice" 9 18 ;;
    2) gdialog --title "Questionnaire" --msgbox "Excellent" 9 18 ;;
    *) gdialog --title "Questionnaire" --checklist "Choose the lanuage you learned"  9 18 5 1 "C" "off" 2 "C++" "off" 3 "Java" "off" 4 "Ruby" "off" 5 "Delpi" "off";;
esac

sleep 1
gdialog --clear
exit 0
```

问题：
（1）分析整个程序的功能层次；
（2）在程序中提取5条新语句，或您认为有用的语句，分析它所对应的功能；
（3）仿照程序，请编写类似功能属于自己开发的小程序。

五、实验结果记录

六、实验结果分析

七、实验心得

Linux 程序设计实验报告 4

——Linux 系统 C 开发工具

一、实验目的

1. 掌握 linux 环境下 C 程序的编辑、编译、运行等操作；
2. 掌握多文件系统的编译及连接库存的生成、应用；
3. 初步掌握 gdb 调度方法；
4. 初步掌握 makefile 工程文件的编写与使用；
5. 掌握系统函数的应用。

二、实验任务与要求

1. 函数库的创建；
2. makefile工程文件的编写；
3. 应用gdb调试程序；
4. 使用gcc编译时连接库的使用。

三、实验工具与准备

PC 机、Linux Redhat Fedora Core6 操作系统。

四、实验步骤与操作指导

任务 1 调试下列程序。

程序通过创建一个小型函数库，演示函数库的创建方法。函数库中包含两个函数，并在一个示例程序中调用其中一个函数。这两个函数分别是 pro1 和 pro2。按下面步骤生成函数库及测试函数库。

步骤 1 为两个函数分别创建各自的源文件（将它们分别命名为 pro1.c 和 pro2.c）。

```
[root@localhost root]# cat pro1.c
#include <sdtio.h>
void pro1(int arg)
```

```
{
    printf("hello: %d\n",arg);
}

[root@localhost root]# cat pro2.c
#include <sdtio.h>
void pro2(char *arg)
{
    printf("您好: %s\n",arg);
}
```

步骤 2 分别编译这两个函数，产生要包含在库文件中的目标文件。这通过调用带有 -c 选项的 gcc 编译器来实现，-c 选项的作用是阻止编译器创建一个完整的程序，gcc 将把源程序编译成目标程序，文件名为以 .o 结尾。如果此时试图创建一个完整的程序将不会成功，因为还未定义 main 函数。

```
[root@localhost root]# gcc -c pro1.c pro2.c
[root@localhost root]# ls *.o
pro1.o pro2.o
```

步骤 3 现在编写一个调用 pro2 函数的程序。首先，为库文件创建一个头文件 lib.h。这个头文件将声明库文件中的函数，它应该被所有希望使用库文件的应用程序所包含。

```
[root@localhost root]# cat lib.h
/*lib.h: pro1.c, pro2.c*/
void pro1(int);
void pro2(char *);
```

步骤 4 主程序（program.c）非常简单。它包含库的头文件并且调用库中的一个函数。

```
[root@localhost root]# cat program.c
#include "lib.h"
int main()
{
    pro2("Linux world");
    exit(0);
}
```

步骤 5 现在，来编译并测试这个程序。暂时为编译器显式指定目标文件，然后要求编译器编译的文件并将其与预先编译好的目标模块 pro2.o 链接。

```
[root@localhost root]# gcc -c program.c
[root@localhost root]# gcc -o program program.o pro2.o
[root@localhost root]# ./program
```

您好：Linux world

步骤 6 现在，创建并使用一个库文件。用 ar 程序创建一个归档文件并将目标文件添加进去。这个程序之所以称为 ar，是因为它将若干单独的文件归并到一个大的文件中以创建归档文件。注意，也可以用 ar 命令来创建任何类型文件的归档文件。

```
[root@localhost root]# ar crv libfoo.a pro1.o pro2.o
```

函数库现在即可使用了。可以在编译器命令行的文件列表中添加该库文件以创建程序，如下所示：

```
[root@localhost root]# gcc -o program program.o libfoo.a
[root@localhost root]# ./program
```

您好：Linux world

也可以用-l 选项来访问函数库，但是因为其未保存在标准位置，所以必须用-L 选项来指示 gcc 在何处可以找到它，如下所示：

```
[root@localhost root]#gcc -o program program.o -L. -lfoo
```

-L.选项指示编译器在当前目录中查找函数库。-lfoo 选项指示编译器使用名为 libfoo.a 的函数库（或者名为 libfoo.so 的共享库，如果它存在的话）。

要查看目标文件、函数库或可执行文件里包含的函数，可使用 nm 命令。如果查看 program 和 libfoo.a，就会看到函数库 libfoo.a 中包含 pro1 和 pro2 两个函数，而 program 里只包含函数 pro2。创建程序时，它只包含函数库中它实际需要的函数。虽然程序中的头文件包含函数库中所有函数的声明，但这并不将整个函数库包含在最终的程序中。

 问题：

（1）按照给出的步骤 1-6 调试程序；

（2）编写两个函数，其中一个求数组中的最大值与最小值，另一函数求某一数的个数，建立一个库，对程序进行调试；

（3）试写出问题（2）中的 makefile 工程文件。

任务 2 调试下列程序。

```c
#include<unistd.h>
int main()
{
char passwd[13];
char *key;
char slat[2];
key= getpass("Input First Password:");
slat[0]=key[0];
slat[1]=key[1];
strcpy(passwd,crypt(key slat));
key=getpass("Input Second Password:");
slat[0]=passwd[0];
slat[1]=passwd[1];
printf("After crypt(),1st passwd :%s\n",passwd);
printf("After crypt(),2nd passwd:%s \n",crypt(key slat));
```

}

程序中需要添加两个预处理命令，另有两个错误。在调试命令中还需要参数-1，请写出调试命令。

任务 3 程序设计题。

在 home/kk 目录下编写一个 a.h 库文件，在 root 目录下编写一个含 main 函数的主程序，在 main 函数中调用在 a.h 中的一个函数。

任务 4 调试下列程序。

调试目的是实现多文件编译。文件结构是：

（1）bubble.h 存放函数 bubble 的声明；
（2）bubble.c 存放函数 bubble 的实现；
（3）main.c 存放 main 函数实现，其中调用了 bubble 函数；
（4）补充完整 makefile 文件。

其中各文件中的代码如下：

```
bubble.h
#ifndef __BUBBLE_H__
#define __BUBBLE_H__

void bubble(int* a, int n);
#endif // __BUBBLE_H__
```

```
bubble.c
#include "bubble.h"
void bubble(int* a, int n)
{
    int i, j, t;
    for(i = 1; i < n; i++)
    {
        for(j = 0; j < n - i; j++)
        {
            if(a[j] > a[j + 1])
            {
                t = a[j];
                a[j] = a[j + 1];
                a[j + 1] = t;
            }
        }
    }
}
```

}

main.c
```c
#include <stdio.h>
#include "bubble.h"

#define COUNT 10
int main(void)
{
    int i;
    int a[COUNT] = {3, 5, 4, 8, 9, 6, 2, 1, 7, 0};
    bubble(a, COUNT);
    for(i = 0; i < COUNT; i++)
    {
            printf("%d ", a[i]);
    }
    printf("\n");

    return 0;
}
```

makefile
```
CC = gcc
FLAGS = -Wall -O2
main: main.o bubble.o
```
（请补充完整makefile文件）

任务5 *程序调试。*

程序的功能是输入两个整数，输出两个整数间的所有自然数。

步骤1 设计编辑源程序代码。

```
[root@localhost root]#vim 4-3.c
```
程序代码如下：
```c
#include <stdio.h>
int num(int x, int y);
int main()
{
  int a1,a2,min_int;
  printf("请输入第一个整数：");
  scanf("%d",&a1);
```

```
    printf("请输入第二个整数：");
    scanf("%d",&a2);
    num(a1,a2);
}
int num(int x, int y)
{
    int temp,i;
    if (x>y)
        {temp=x;x=y;y=temp;}
    for(i=x;i<=y;i++)
        printf("%5d",i);
    printf("\n");
    return 0;
}
```

步骤 2 用 gcc 编译程序。

在编译的时候要加上选项"-g"。这样编译出的可执行代码中才包含调试信息，否则之后 gdb 无法载入该可执行文件，请写出编译命令。

[root@localhost root]#

步骤 3 进入 gdb 调试环境。

gdb 进行调试的是可执行文件，因此要调试的是 4-3 而不是 4-3.c，输入如下，请输入 gdb 的调试命令：

[root@localhost root]#

回车后就进入了 gdb 调试模式。

```
GNU gdb Red Hat Linux (5.3post-0.20021129.18rh)
Copyright 2003 Free Software Foundation, Inc.
GDB is free software, covered by the GNU General Public License, and you are
welcome to change it and/or distribute copies of it under certain conditions.
Type "show copying" to see the conditions.
There is absolutely no warranty for GDB.   Type "show warranty" for details.
This GDB was configured as "i386-Redhat-Linux-gnu"...
(gdb)
```

图 4.1 进入 gdb 调试环境

可以看到，在 gdb 的启动画面中有 gdb 的版本号、使用的库文件等信息，在 gdb 的调试环境中，提示符是"(gdb)"。

步骤 4 用 gdb 调试程序。

（1）查看源文件

在 gdb 中输入"1"(list)就可以查看程序源代码，一次显示 10 行。

可以看出，gdb 列出的源代码中明确地给出了对应的行号，这样可以大大方便代码的定位。

（2）设置断点

设置断点在调试程序时是一个非常重要的手段，它可以使程序到一定位置暂停运行。软件工程师可以在断点处查看变量的值、堆栈情况等，从而找出代码的问题所在。

在 gdb 中设置断点命令是"b"(break)，后面跟行号或者函数名。

不指定具体行号的断点设置在"b"(break)后面跟函数名。本例可以输入"break num"。

（3）查看断点信息

设置完断点后，可以用命令"info b"(info break)查看断点信息。

（4）运行程序

接下来可以运行程序，可以输入"r"(run)开始运行程序。

（5）查看与设置变量值

调试程序重要手段就是查看断点处的变量值。程序运行到断点处会自动暂停，此时输入"p 变量名"可查看指定变量的值。

调试程序时，若需要修改变量值，可程序运行到断点处时，输入"set 变量=设定值"，例如，给变量"a2" 赋值 100，输入"set a2=100"。

gdb 在显示变量值时都会在对应值前加"$n"标记，它是当前变量值的引用标记，以后想再引用此变量，可以直接使用"$n"，提高了调试效率。

（6）单步运行

很多情况下，调试的时候要单步运行程序。在断点处输入 "n"(next)或者"s"(step)可单步运行。它们之间的区别在于：若有函数，调用时，"s"会进入该函数，而"n"不会进入该函数。

（7）继续运行程序

在查看完变量或堆栈情况后可以输入"c"(continue)命令恢复程序的正常运行，把剩余的程序执行完，并显示执行结果。

（8）退出 gdb 环境

退出 gdb 环境只要输入"q"(quit)命令，回车后退出 gdb 环境。

五、实验结果记录

六、实验结果分析

七、实验心得

Linux 程序设计实验报告 5

——Linux 环境系统函数的应用

一、实验目的

1. 掌握随机数函数的使用方法；
2. 掌握结构体 struct timeval 的成员 tv_sec 与 tv_usec 的应用；
3. 掌握时间函数 time、localtime、gettimeofday 的使用；
4. 掌握系统函数 system、tcgetattr 等的应用。

二、实验任务与要求

1. 随机数函数的使用；
2. 猜数游戏的程序；
3. 结构体 struct timeval 的成员 tv_sec 与 tv_usec 的应用；
4. 时间函数在简单记事本程序中的应用；
5. 时间函数在计时器，有暂停、查看、重置程序设计设计中的应用。

三、实验工具与准备

PC 机、Linux Redhat Fedora-Core6 操作系统。

四、实验步骤与操作指导

任务 1 调试下列程序。

产生 10 个介于 1~10 之间的随机数值。提示函数 rand( ) 会返回一个 0~ RAND_MAX（其值为 2147483647）之间的随机值。产生一个大于等于 0、小于 1 的数，此数可表示为：rand()/(RAND_MAX+1.0)。

```
[root@localhost root]#vim 5-1.c
```
程序代码如下：
```
#include <stdlib.h>
#include "stdio.h"
```

```
int main()
{
 int i,j;
 srand((int)time(0));
 for(i=0;i<10;i++)
 {
  j=1+(int)(10.0*rand()/(RAND_MAX+1.0));
  printf(" %d ",j);
 }
 printf("\n");
}
```

问题：

（1）论述语句 srand((int)time(0)); 的功能；

（2）修改程序，产生 50 个 100~1000 的随机整数。

任务 2 程序设计。

编写一个猜数游戏的程序，先产生一个随机数，要求被试输入一个数，计算机会提示猜大了，猜小了或恭喜您猜中了，直到猜中，退出程序。修改程序，限定猜数的次数作为难度系数，除了提示猜大了，猜小了或恭喜您猜中了外，还有次数已到，猜数失败。

任务 3 调试下列程序。

应用结构体 struct timeval 的成员 tv_sec 与 tv_usec 显示系统时间的秒与微秒，并测试输出成员 tv_sec 与 tv_usec 这段程序所用时间。提示程序设计的步骤为用函数 gettimeofday 读取系统时间，显示系统中的秒与微妙，显示 Greenwich 的时间差，测试系统时间 tvz，计算时间差。

步骤 1 编辑源程序代码。

```
[root@localhost root]#vim 5-2.c
```

程序代码如下：

```c
#include<sys/time.h>
#include<unistd.h>
int main()
{
struct timeval tv1,tv2;
struct timezone tz;
gettimeofday(&tv1,&tz);

printf("tv_sec; %d\n", tv1.tv_sec) ;
```

```
        printf("tv_usec; %d\n",tv1.tv_usec);
        gettimeofday(&tv2,&tz);
        printf("tv2_usec-tv1_usec; %d\n",tv2.tv_usec-tv1.tv_usec);
        return 0;
    }
```

 问题：

（1）修改以上程序，分别编写一个有参数的函数及宏定义，研究它们的调用效率问题；

（2）编写一个程序，如查找或排序效率问题，查找算法有顺序、二分查找等。

任务 4 编写程序。

编写一个简单的记事本，其功能是实现输入任务及任务截止时间，通过系统时间和日期函数的使用，可计算输出任务的剩余时间。

 提示：

```
printf("请输入任务截止时间，分别输入日期1~31，小时0~23、分、秒对应的数字\n");
scanf("%d%d%d%d", &i,&j,&k,&l);
```

为对应的输入时间，即余下时间可以从以下语句中得到。

```
time (&timep);
p=localtime(&timep);
a=(i-p->tm_mday)*24*3600+(j-p->tm_hour)*3600+(k-p->tm_min)*60+l-p->tm_sec;
```

在余下时间为0时可以考虑响铃及显示任务。

任务 5 调试下列程序。

```
#include<stdio.h>
#include<stdlib.h>
#include<sys/time.h>
#include<unistd.h>
#include<termios.h>

int mygetch()              //实时返回键盘被按下的值
{
    struct termios oldt, newt;
    int ch;
    tcgetattr(STDIN_FILENO,&oldt);
    newt = oldt;
    newt.c_lflag &= ~(ICANON|ECHO);
    tcsetattr(STDIN_FILENO,TCSANOW,&newt);
```

```c
        ch = getchar();
        tcsetattr(STDIN_FILENO,TCSANOW,&oldt);
        return ch;
    }

    int main()
    {
        char c, v;
        int i, rn = 0, n;
        struct timeval tv1, tv2;
        struct timezone tz;
        double sum = 0, prob = 0;
        srand((unsigned)time(NULL));
        printf("Please input the number of tests:");
        scanf("%d",&n);
        getchar();
        while(n)
        {
           for(i = 0; i < n; i++)
           {
              c = (char)(1.0+(int)(25.0*rand()/RAND_MAX)+65.0);
              printf("\033[5;5H%c",c);
              sleep(0.5);
              gettimeofday(&tv1,&tz);
              system("clear");
              v = mygetch();
              gettimeofday(&tv2,&tz);
              if(v == c)
               {
                  rn++;
    sum+=(tv2.tv_sec-tv1.tv_sec)+(tv2.tv_usec-tv1.tv_usec)/1000000.0;
              }
           }
        prob = rn / n;
        system("clear");
        if(prob == 0)          //全都按错时鼓励下并提醒是不是忘记打开大写开关
            printf("sorry, you score 0, please try again~~, maybe you forgot to switch the Caps Lock on\n");
        else
```

```
            printf("the accuracy is %lf, and the average time of response is %lf secs\n",prob,sum/rn);
        printf("Please input the number of tests(input 0 to quit):");
        scanf("%d",&n);
        getchar();
        rn = 0;
        sum = 0;  //把 rn 和 sum 清零
        }


}
```

问题：
（1）调试程序，分析并画出程序执行的流程图；
（2）修改程序，编写界面更为优美的测试程序。

任务 6　调试下列程序。
程序的功能是计时器，有暂停、查看、重置等功能。

```
#include<stdio.h>
#include<stdlib.h>
#include<sys/time.h>
#include<unistd.h>
int main()
{
long int begin,sec,stop;
struct timeval tv1, tv2;
struct timezone tz;
char tmp;
begin=0;
stop=0;
sec=0;
system("clear");
printf("计时器程序（单位 s）\n");
printf("输入 b(begin)计时器开始计时\n");
printf("输入 w(watch)查看已经累计时间\n");
printf("输入 r(rest)重新开始计时\n");
printf("输入 s(stop)暂停计时器\n");
printf("输入 e(end)结束计时器\n");
while(1)
```

```c
{
scanf("%c",&tmp);

if(tmp=='b'){
    if(begin==1&&stop==0) printf("计时器已经启动!\n");
    if(begin==0&&stop==0)  {
                printf("计时器启动\n");
                gettimeofday(&tv1,&tz);
                sec=0;
                begin=1;
        }
    if(stop==1){
            gettimeofday(&tv1,&tz);
            stop=0;
            printf("暂停结束!\n");
            }

}
if(tmp=='w'){
    if(stop==0){
            gettimeofday(&tv2,&tz);
            printf("已经计时%ld 秒\n",sec+tv2.tv_sec-tv1.tv_sec);
            }
    if(stop==1)
            printf("已经计时%ld 秒\n",sec);
}
if(tmp=='s'){
    if(stop==1){
            printf("计时已经暂停!\n");
            }

if(stop==0){
            gettimeofday(&tv2,&tz);
            sec=sec+tv2.tv_sec-tv1.tv_sec;
            printf("计时暂停,已经计时%ld 秒\n",sec);
            stop=1;
            }


}
```

```c
        if(tmp=='r'){
            gettimeofday(&tv2,&tz);
            printf("已经计时%ld 秒\n",sec+tv2.tv_sec-tv1.tv_sec);
            printf("计时器在 5 秒后被重置！\n");
            sleep(5);
            begin=0;
            sec=0;
            stop=0;
            system("clear");
            printf("计时器程序（单位 s）\n");
            printf("输入 b(begin)计时器开始计时\n");
            printf("输入 w(watch)查看已经累计时间\n");
            printf("输入 r(rest)重新开始计时\n");
            printf("输入 s(stop)暂停计时器\n");
            printf("输入 e(end)结束计时器\n");
        }
        if(tmp=='e') break;
    }
    return 0;
}
```

问题：

（1）写出程序中起计时器功能的语句；

（2）写出计时器暂停功能的语句；

（3）写出重置计时器的语句。

五、实验结果记录

六、实验结果分析

七、实验心得

Linux 程序设计实验报告 6

——Linux 文件 I/O 操作 1

一、实验目的

1. 掌握函数 stat 中文件属性的应用；
2. 掌握系统函数 system、opendir、scandir 的使用；
3. 初步掌握 struct dirent 的结构体变量的使用方法；
4. 掌握文件阻塞与非阻塞 I/O 的操作；
5. 掌握文件属性的判断。

二、实验任务与要求

1. 测试文件S_IRUSR、S_IWUSR、S_IRGRP、S_IROTH属性；
2. 应用system函数对网站的网络连通情况进行测试；
3. 应用readdir函数显示文件和子目录；
4. 文件属性的判断；
5. 阻塞 I/O 文件操作的程序设计。

三、实验工具与准备

PC 机、Linux Redhat Fedora Core6 操作系统。

四、实验步骤与操作指导

任务 1 程序设计。
程序应用 system 函数建立/home/liu 文件，应用 chmod 函数使文件 liu 具有 S_IRUSR、S_IWUSR、S_IRGRP、S_IROTH 属性，最后应用函数 stat 获取文件的大小与建立的时间。

任务 2 程序设计。
编写程序在程序中能够对 yahoo 网站的网络连通情况进行测试。

任务3 调试下列程序。

程序的功能是要求打印系统目录"/etc/rc.d"下所有的文件和子目录的名字。

程序代码如下：

```
#include<stdio.h>
#include<sys/types.h>
#include<dirent.h>
#include<unistd.h>
int main()
{
  DIR * dir;
  struct dirent * ptr;
  dir=opendir("/etc/rc.d");
  printf("/etc/rc.d目录中文件或子目录有:\n");
  while((ptr = readdir(dir))!=NULL)
  {
    printf("%s\n",ptr->d_name);
  }
  closedir(dir);
}
```

问题：

修改程序，要求读取"/etc"目录下所有的目录结构，并依字母顺序排列。

　　#include<dirent.h>

考虑以下语句：

　　scandir("/etc",&namelist,0,alphasort);

程序段：

```
while(n--)
{
    printf("%s\n", namelist[n]->d_name);
    free(namelist[n]);
}
```

任务4 调试并分析下列程序的结果。

程序的功能是用递归的方法列出某一目录下的全部文件的大小和文件夹及创建日期，包括子文件和子文件夹。

程序代码如下：

```
#include<stdio.h>
#include<time.h>
```

207

```c
#include<linux/types.h>
#include<dirent.h>
#include<sys/stat.h>
#include<unistd.h>
#include<string.h>
char *wday[]={"日","一","二","三","四","五","六"};
void list(char *name,int suojin)
{
    DIR *dirname;
    struct dirent *content;
    struct stat sb;
    struct tm *ctime;
    int i;
    if((dirname=opendir(name))==NULL)
    {
         printf("该目录不存在\n");
          return;
    }
    chdir(name);/*改换工作目录*/
    while((content=readdir(dirname))!=NULL)
    {


     for(i=0;i<suojin;i++)
         putchar('\t');
      if(content->d_type==4)
          printf("目录\t");
      else if(content->d_type==8)
          printf("文件\t");
      else
      printf("其他\t");
      stat(content->d_name,&sb);
      ctime=gmtime(&sb.st_mtime);
   printf("%d 年%d 月%d 日 星期%s %d:%d:%d\t",ctime->tm_year+1900,
 1+ctime->tm_mon,ctime->tm_mday,wday[ctime->tm_wday],ctime->tm_hour,
ctime->tm_min,ctime->tm_sec);
    printf("%d\t",sb.st_size);
    printf("%s\n",content->d_name);/*列出目录或文件的相关信息*/
    if(content->d_type==4&&strcmp(content->d_name,"..")&&strcmp(content->
```

```
    d_name,"."))
        {
                list(content->d_name,suojin+1);    /*如果是目录,则递归列出目录里的内容*/
            }
        }
            closedir(dirname);
            chdir("..");       /*当该层目录中的文件列完后,返回父目录*/
    }


    int main(int argc,char *argv[])
    {
      char name[256];
      printf("类型\t 最后修改时间\t\t\t 大小\t 文件名\n");
      printf("******************************************************\n");
      if(argc==1) {
            printf("Enter directory name:");
            scanf("%s",name);
            list(name,0);
        }
        else
        {
            list(argv[1],0);
        }
    }
```

任务5 程序设计题。

设计一个程序,要求判断"/etc/passwd"的文件类型。提示:使用 st_mode 属性,可以使用几个宏来判断:S_ISLNK(st_mode) 是否是一个连接,S_ISREG 是否是一个常规文件 S_ISDIR 是否是一个目录,S_ISCHR 是否是一个字符设备,S_ISBLK 是否是一个块设备,S_ISFIFO 是否是一个 FIFO 文件,S_ISSOCK 是否是一个 SOCKET 文件。

任务6 调试下列程序。

程序主要运用一个非阻塞的 I/O 操作,运行时首先打开当前终端文件/dev/tty,并指定 O_NONBLOCK 标志。程序运行时每隔一定时间(6 秒)等待用户从终端输入,总共等待 30 秒,每次等待时屏幕都有提示 "try again",30 秒后程序继续执行主程序,输出以下图形后结束。

        *
       * *

```
        * *  *
        * *  * *
        * *  * * *
#include <unistd.h>
#include <fcntl.h>
#include <errno.h>
#include <string.h>
#include <stdlib.h>

#define MSG_TRY "try again\n"
#define MSG_TIMEOUT "timeout\n"

int main(void)
{
    char buf[10];
    int fd, n, i, j;
    fd = open("/dev/tty", O_RDONLY|O_NONBLOCK);
    if(fd<0) {
            perror("open /dev/tty");
            exit(1);
    }
    for(i=0; i<5; i++) {
            n = read(fd, buf, 10);
            if(n>=0)
                    break;
            if(errno!=EAGAIN) {
                    perror("read /dev/tty");
                    exit(1);
            }
            sleep(6);
            write(STDOUT_FILENO, MSG_TRY, strlen(MSG_TRY));
    }
    if(i==5)
            write(STDOUT_FILENO, MSG_TIMEOUT, strlen(MSG_TIMEOUT));
    else
            write(STDOUT_FILENO, buf, n);
        for(i=0;i<5;i++){
            for(j=0;j<=i;j++)
                printf("%2c",'*');
```

```
            printf("\n");
        }
        close(fd);
        return 0;
    }
```

程序运行时，轮询等待用户的输入，等待期间如有输入，即转入主程序执行，如没有输入，30 秒后执行主程序。

  问题：

（1）修改程序，程序性质为阻塞的；

（2）修改程序，程序运行中不断地向文件 liu 写入 ps 的情况，如果有输入，则测试与 yahoo 的连通情况。

五、实验结果记录

六、实验结果分析

七、实验心得

Linux 程序设计实验报告 7

——Linux 文件 I/O 操作 2

一、实验目的

1. 掌握文件锁的使用；
2. 掌握文件目录与文件的递归及深度遍历；
3. 掌握文件属性的判断。

二、实验任务与要求

1. 文件加锁、解锁的操作，文件锁属性的判断；
2. 应用函数 opendir、readdir 显示文件与文件目录；
3. 在指定的不同目录中，探索相同的文件；
4. 应用函数 readdir 及文件属性编写类似于命令"ls -l"的程序设计。

三、实验工具与准备

PC 机、Linux Redhat Fedora Core6 操作系统。

四、实验步骤与操作指导

任务 1 调试下列程序。

程序的功能是在"/root"下打开一个名为"7-1file"的文件，如果该文件不存在，则创建此文件。打开后对其加上强制性的写入锁 F_WRLCK，按回车后解锁 F_UNLCK，然后加上读出锁 F_RDLCK，按回车后再解锁 F_UNLCK。程序在终端1运行后会显示程序的进程号，再打开终端 2，会提示此文件处于锁定状态，此时在终端 2 可以多按回车，观察程序的运行结果。然后在终端 1 按回车，等待终端 1 解锁后，在终端 2 才可锁定此文件，你可观察到强制性锁是独占状态，当在终端 2 解锁后，在终端 1 或 2 可加读出锁，在读出锁状态终端 1 或 2 的运行不需要等待，因为读出锁是处于共享状态，请编写程序并测试程序运行的结果。主程序先用 open 函数打开文件"7-1file"，如果该文件不存在，则创建此文件；接着调用自定义函数 lock_set：先传递参数"F_WRLCK"给文件"7-1file"加锁，并打印输出给文件加锁进程的进程号,然后先传递参数"F_UNLCK"给文件"7-1file"解锁,并打印输出给文件解锁进程的进程号;在自定义函数 lock_set 给文件上锁语句前,加上判断文件是否上锁的语句,如果文件已经被上锁,打印输出给文件上锁进程的进程号。

程序代码如下:

```c
/*7-1.c 程序:打开"/home/7-1file"后对其加上强制性的写入锁,然后释放写入锁*/
#include<stdio.h>
#include<stdlib.h>
#include <unistd.h>
#include <sys/file.h>
#include <sys/types.h>
#include <sys/stat.h>
void lock_set(int fd, int type)
{
  struct flock lock;
  lock.l_whence = SEEK_SET;
  lock.l_start = 0;
  lock.l_len =0;
  while(1)
  {
    lock.l_type = type;
    if((fcntl(fd,F_SETLK,&lock))==0)
    {/*根据不同的 type 值给文件加锁或解锁*/
      if( lock.l_type == F_RDLCK )   /*F_RDLCK 为共享锁,表示读取锁或建议性锁*/
        printf("加上读取锁的是: %d\n",getpid());
      else if( lock.l_type == F_WRLCK )   /*F_WRLCK 为排斥锁,表示强制性锁*/
        printf("加上写入锁的是: %d\n",getpid());
      else if( lock.l_type == F_UNLCK )
        printf("释放强制性锁: %d\n",getpid());
    return;
    }
    fcntl(fd, F_GETLK,&lock); /*读取文件锁的状态*/
    if(lock.l_type != F_UNLCK)
    {
    if( lock.l_type == F_RDLCK )
      printf("文件已经加上了读取锁,其进程号是: %d\n",lock.l_pid);
    else if( lock.l_type == F_WRLCK )
      printf("文件已加上写入锁,其进程号是: %d\n",lock.l_pid);
    getchar();
```

```
        }
     }
}
int main ()
{
int fd;
fd=open("/root/7-1file",O_RDWR | O_CREAT, 0666);
if(fd < 0)
{
perror("打开出错");
exit(1);
}
lock_set(fd, F_WRLCK);
getchar();
lock_set(fd, F_UNLCK);
getchar();
lock_set(fd, F_RDLCK);
getchar();
lock_set(fd, F_UNLCK);
close(fd);
exit(0);
}
```

问题：

（1）在不同的终端同时运行该程序，分析程序的运行情况；

（2）修改程序，如果把文件名作一个房号，请描述宾馆客人入住的情况。

任务2 程序调试。

调试下列程序，程序应用目录函数，通过递归函数进行深度遍历，达到显示输入目录文件下的所有文件，包括子目录下的所有文件，并统计文件数量。

```
#include<dirent.h>
#include<stdio.h>
#include<stdlib.h>
#include<sys/stat.h>
#include<sys/types.h>
#include<unistd.h>
#include<string.h>
#define MAXFILE 100          /*定义最大文件名长度,此程序暂定100*/
```

```c
int num=0;
/***************************************************************
函数功能：遍历整个目录，并递归进入其子目录，实现深度遍历
参数：含有目录路径的指针
***************************************************************/
void viewdir(char *pfilename)
{
DIR *dir;
char file[MAXFILE];
char filename[MAXFILE];
strcpy(filename, pfilename);
struct dirent *ptr;
struct stat buf;
dir = opendir(filename);
if(dir == NULL)
return;
strcpy(file,filename);
while((ptr = readdir(dir))!=NULL)
{
    strcat(filename,"/");
    strcat(filename,ptr->d_name);
    stat(filename,&buf);
    /*下行判断待定文件是否为目录文件，且是否为.和..*/
    printf("%s\n",filename);
    /*若均非，则对其运行viewdir函数*/
    if((S_ISDIR(buf.st_mode)==1)&&(strcmp((ptr->d_name),".")!=0)&&
                (strcmp((ptr->d_name),"..")!=0))
    {
        num++;
        viewdir(filename);                    /*调用递归*/
    }
    else  if(S_ISDIR(buf.st_mode)==0)
        num++;
    trcpy(filename,file);
}
closedir(dir);
}

int main()
```

```c
{
    char filename[MAXFILE];
    struct stat buf;
    printf("请输入目录路径:");
    scanf("%s",filename);
    stat(filename,&buf);
    if(S_ISDIR(buf.st_mode)==0)
    /*检验指定文件是否为目录文件*/
    {
        printf("此文件非目录文件！\n");
        exit(0);
    }
    viewdir(filename);
    printf("该目录中的文件数量为：%d\n",num);
    return 0;
}
```

问题：

（1）如何使用 md5sum 命令把某文件报文摘要输出到一个 md5 文件；
（2）改写程序，采用主函数调用比较函数。

任务3 程序调试。

程序的功能是通过使用命令 md5sum 计算路径下文件的 md5 值，并找出所有相同的重复冗余文件并删除，达到节省磁盘空间的目的，相当于一个文件冗余比较器。程序运行后输入两个不同的路径，比较这两个路径下的文件差异。

```c
#include<stdio.h>
#include<stdlib.h>
#include<dirent.h>
#include<sys/types.h>
#include<sys/stat.h>
#include<fcntl.h>
#include<string.h>

int main()
{
    int same = 0, del = 0, node = 0, i;
    char md5[10000][40], temp[40], comm[1000];
    char dir1[1000], dir2[1000];
```

```c
    FILE *fp;
    DIR *dirname;
    struct dirent *content;
printf("==========================\n");
    printf("    欢迎使用文件比较器\n");
    printf("==========================\n");
    printf("请输入要比较的路径：\n");
    scanf("%s", &dir1);
    scanf("%s", &dir2);
    printf("正在比较...");
    if ((dirname = opendir(dir1)) == NULL){
      printf("%s 该目录不存在\n", dir1);
      return 0;
    }
    chdir(dir1);
    while((content = readdir(dirname)) != NULL){
      if (content -> d_type == 8){
        strcpy(comm, "md5sum ");
        strcat(comm, content -> d_name);
        fp = popen(comm, "r");
        fgets(temp, 32, fp);
        for (i = 0; i <= node - 1; i++)
          if (strcmp(temp, md5[i]) == 0){
            same++;
            strcpy(comm, "rm ");
            strcat(comm, content -> d_name);
            if (system(comm) == 0) del++;
            break;
          }
        if (i == node){
          node++;
          strcpy(md5[node - 1], temp);
        }
      }
    }
    close(dirname);
    if ((dirname = opendir(dir2)) == NULL){
      printf("%s 该目录不存在\n", dir2);
      return 0;
```

```
        }
        chdir(dir2);
        while((content = readdir(dirname)) != NULL){
          if (content -> d_type == 8){
            strcpy(comm, "md5sum ");
            strcat(comm, content -> d_name);
            fp = popen(comm, "r");
            fgets(temp, 32, fp);
            for (i = 0; i <= node - 1; i++)
              if (strcmp(temp, md5[i]) == 0){
                same++;
                strcpy(comm, "rm ");
                strcat(comm, content -> d_name);
                if (system(comm) == 0) del++;
                break;
              }
            if (i == node){
              node++;
              strcpy(md5[node - 1], temp);
            }
          }
        }
        close(dirname);
        printf("\n");
        printf("比较结束\n");
        printf("共发现%d 个相同的文件,已删除%d 个冗余文件\n", same, del);
        return 0;
    }
```

问题:

(1) 如何使用 md5sum 命令把某文件报文摘要输出到一个 md5 文件;
(2) 改写程序,采用主函数调用比较函数。

任务 4 设计一个程序,要求根据输入的文件或目录(若缺省则为当前目录),实现命令 "ls -l" 的功能,不使用 system 函数。

```
#include<stdio.h>
#include<stdlib.h>
#include<string.h>
```

```c
#include<unistd.h>
#include<fcntl.h>
#include<sys/stat.h>
#include<dirent.h>
#include <pwd.h>
#include <grp.h>
void op_l(struct dirent *dip,struct stat buf);
void op_l2(char *ar,struct stat buf);
int main(int argc,char* argv[])
{
    char op[100];
    char tmp[100];
    DIR *dp;
    int fd;
    struct stat buf;
    struct dirent *dip;
    switch(argc)                          //判断命令行参数的缺省情况
    {
        case 1:                                       // 没有对象
            dp=opendir("./");
            while ((dip=readdir(dp))!=NULL) {
                if( dip->d_name[0]=='.')
                    continue;
                strcpy(tmp,".");
                strcat(tmp, "/");
                strcat(tmp,dip->d_name);
                stat(tmp,&buf);
                op_l(dip,buf);
            }
            break;
        case 2:
            if((dp=opendir(argv[1]))!=NULL){          // 对象为目录
                while ((dip=readdir(dp))!=NULL) {
                    if( dip->d_name[0]=='.')
                        continue;
                    strcpy(tmp,argv[1]);
                    strcat(tmp, "/");
                    strcat(tmp,dip->d_name);
                    stat(tmp,&buf);
```

```c
                    op_l(dip,buf);
            }
        }
        else{            // 对象为文件
            stat(argv[1],&buf);
            op_l2(argv[1],buf);
        }

        break;
    }
    return 0;
}
void op_l(struct dirent *dip,struct stat buf)        // 关于ls -l命令的函数
{
    char per[11];
    unsigned int mask=0700;
    struct passwd *uname;
    struct group *gname;
    char *date;
    int N_BITS = 3;
    int i=3;
    int j=0;
    static char* permess[]={"---","--x","-w-","-wx","r--","r-x", "rw-","rwx"};
    if(S_ISREG(buf.st_mode))                          //文件类型
        per[0]='-';
    else if(S_ISDIR(buf.st_mode))
        per[0]='d';
    else if(S_ISCHR(buf.st_mode))
        per[0]='c';
    else if(S_ISBLK(buf.st_mode))
        per[0]='b';
    else if(S_ISFIFO(buf.st_mode))
        per[0]='p';
    else if(S_ISLNK(buf.st_mode))
        per[0]='l';
    else if(S_ISLNK(buf.st_mode))
        per[0]='s';
    else
        per[0]='*';
```

```c
        while(i>0) {                                              // 权限
            per[1+j*3]=permess[(buf.st_mode & mask)>>(i-1)*N_BITS][0];
            per[2+j*3]=permess[(buf.st_mode & mask)>>(i-1)*N_BITS][1];
            per[3+j*3]=permess[(buf.st_mode & mask)>>(i-1)*N_BITS][2];
            i--;
            j++;
            mask>>=N_BITS;
        }
        per[10]='\0';
        printf("%s ",per);
        printf("%d ",buf.st_nlink);                        //硬链接数
        uname = getpwuid(buf.st_uid);                      //用户id转变为用户名
        gname = getgrgid(buf.st_gid);                      //组id转变为组名
        printf("%-10s%-10s",uname->pw_name,gname->gr_name);
        printf("%-6u",(unsigned int)buf.st_size);          //文件大小
        date = (char*)ctime(&buf.st_mtime);                //最后修改时间
        date[24]='\0';
        printf("%-10s  %-10s\n",date,dip->d_name);
}
void op_l2(char *ar,struct stat buf)          // 关于-l命令的函数（对单个文件）
{
        char per[11];
        unsigned int mask=0700;
        struct passwd *uname;
        struct group *gname;
        char *date;
        int N_BITS = 3;
        int i=3;
        int j=0;
        static                                                                char*
permess[]={"---","--x","-w-","-wx","r--","r-x","rw-","rwx"};

        if(S_ISREG(buf.st_mode))                           //文件类型
            per[0]='-';
        else if(S_ISDIR(buf.st_mode))
            per[0]='d';
        else if(S_ISCHR(buf.st_mode))
            per[0]='c';
        else if(S_ISBLK(buf.st_mode))
```

```c
        per[0]='b';
    else if(S_ISFIFO(buf.st_mode))
        per[0]='p';
    else if(S_ISLNK(buf.st_mode))
        per[0]='l';
    else if(S_ISLNK(buf.st_mode))
        per[0]='s';
    else
        per[0]='*';
    while(i>0) {                                              // 权限
        per[1+j*3]=permess[(buf.st_mode & mask)>>(i-1)*N_BITS][0];
        per[2+j*3]=permess[(buf.st_mode & mask)>>(i-1)*N_BITS][1];
        per[3+j*3]=permess[(buf.st_mode & mask)>>(i-1)*N_BITS][2];
        i--;
        j++;
        mask>>=N_BITS;
    }
    per[10]='\0';
    printf("%s ",per);
    printf("%d ",buf.st_nlink);
    uname = getpwuid(buf.st_uid);
    gname = getgrgid(buf.st_gid);
    printf("%-10s%-10s",uname->pw_name,gname->gr_name);
    printf("%-6u",(unsigned int)buf.st_size);
    date = (char*)ctime(&buf.st_mtime);
    date[24]='\0';
    printf("%-10s  %-10s\n",date,ar);
}
```

五、实验结果记录

六、实验结果分析

七、实验心得

Linux 程序设计实验报告 8

——进程控制

一、实验目的

1. 掌握进程的常用终端命令；
2. 掌握用 system、exec 函数簇、fork 函数创建进程；
3. 掌握 waitpid 函数的应用；
4. 掌握守护进程编程过程；
5. 掌握守护进程在各种监控中的应用。

二、实验内容

1. 进程的常用终端命令；
2. 用 execl 函数创造进程；
3. 应用 fork 函数创建子进程；
4. 应用 fork 函数创建子进程,在父子进程中执行不同的任务；
5. waitpid 函数的应用；
6. 守护进程的编写与应用；
7. 守护进程在邮件监控中的应用；
8. 守护进程在文件监视中的应用。

三、实验设备与实验准备

PC 机、Linux Redhat Fedora Core6 操作系统。

四、实验步骤与操作指导

任务 1

（1）学习 at 指令的使用，写出 at 指令的使用格式。在当前时间 2 分钟后，通过 at 指令运行命令"ls -l"

（2）使用命令 kill，显示 Linux 环境下的信号。分析这些信号的特点。

(3) 下列是进程调度中常用的函数，通过网络资源查找这些函数的功能与用法。

函 数	功能与用法
kill	
raise	
alarm	
pause	
signal	
sigemptyset	
sigfillset	
sigaddset	
sigdelset	
sigismember	
sigprocmask	

任务 2 程序设计。

用 execl 函数创造进程 ls -l，用 execvp 函数创造进程 ps -ef。

提示：

显示当前目录下的文件信息可执行以下语句：

(1) execl("/bin/ls", "ls", "-al", NULL)；

(2) char *arg[] = {"ps", "-ef", NULL}；
    execvp("ps", arg)。

任务 3 调试下列程序，改正程序中少量的错误，写出程序的功能与程序的运行结果。

```
#include<stdio.h>              /*文件预处理,包含标准输入输出库*/
#include<stdlib.h>             /*文件预处理,包含system、exit等函数库*/
#include<signal.h>             /*文件预处理,包含kill、raise等函数库*/
#include<sys/types.h>   /*文件预处理,包含waitpid、kill、raise等函数库*/
#include<sys/wait.h>           /*文件预处理,包含waitpid函数库*/
#include<unistd.h>     /*文件预处理,包含进程控制函数库*/
int main ()            /*C程序的主函数,开始入口*/
{
pid_t result;
int ret;
result=fork();         /*调用fork函数,复制进程,返回值存在变量result中*/
int newret;
if(result<0)   /*通过result的值来判断fork函数的返回情况,这儿进行出错处理*/
```

```
{
    perror("创建子进程失败");
    exit(1);
}
else if (result==0)      /*返回值为0代表子进程*/
{
    raise(SIGSTOP);  /*调用raise函数,发送SIGSTOP 使子进程暂停*/
    exit(0);
}
else                     /*返回值大于0代表父进程*/
{
    printf("子进程的进程号(PID)是:%d\n",result);
    if((waitpid(NULL,WNOHANG))==0)
      {
        if(ret=kill(result,SIGKILL)!=0)
        /*调用kill函数,发送SIGKILL信号结束子进程 result */
         printf("用kill函数返回值是:%d,发出的SIGKILL信号结束的进程进程号:%d\n",ret,result);
        else{ perror("kill函数结束子进程失败");}
      }
}
}
```

改写程序,父进程等待较长时间在作业的子进程,子进程退出后,父进程再退出。

任务4 *程序设计*。

(1) 在父子进程中分别编写循环程序,应用函数 sleep 的不同参数等,体现程序中父子进程的并发运行。

(2) 在父子进程中分别执行不同的任务,例如在子进程中执行文件编辑任务,在父进程中执行网络连通情况的测试任务,子进程退出后父进程才退出。

任务5 *调试下列程序*。

程序的功能是如何运用 wait 函数避免子进程成为僵尸进程。

```
#include<stdio.h>
#include<unistd.h>
#include<sys/types.h>
#include<sys/wait.h>
int main ()
{
pid_t pid,wpid;
```

```
    int status,i;
    pid=fork();
    if(pid==0)
    {
        printf("这是子进程,进程号(pid)是:%d\n",getpid());
        sleep(5);              /*子进程等待5秒钟*/
        exit(6);
    }
    else
    {
        printf("这是父进程,正在等待子进程……\n");
        wpid=wait(&status);    /*父进程调用wait函数,消除僵尸进程*/
        i=WEXITSTATUS(status);
        printf("等待的进程的进程号(pid)是:%d ,结束状态:%d\n",wpid,i);
    }
}
```

问题:

改写程序,在子程序中用函数 system 启动一个较长时间运行的任务,而在父进程中执行完成任务后,应用 waitpid 函数等待子进程,子进程退出后父进程才退出。

任务 6 程序调试。

程序功能是设计一个闹钟程序,用户输入时间,格式为小时:分钟,例如 9:18 表示设定的时间为9时18分,到了设定时间后,发出蜂鸣声作为提示音,为保证检测时间的准确性,要求使用守护进程。

```
#include <stdio.h>
#include <time.h>
#include <unistd.h>
#include <signal.h>
#include <sys/param.h>
#include <sys/types.h>
#include <sys/stat.h>
void init_daemon(void);
int main()
{
    int hour,min;
    time_t timep;
    struct tm *p;
```

```c
        time (&timep);
        p=localtime(&timep);/*获取系统时间*/
        printf("这是一个闹钟程序,输入你想要设定的时间:\n");
        scanf("%d:%d",&hour,&min);
        init_daemon();  /*调用守护进程*/
        while(1)
        {
                sleep(20);/*每隔20秒检查一下时间是否已到*/
                if(p->tm_hour==hour &&p->tm_min==min)
                {
                        printf("时间到了!\n");
                        printf("\7\7\7\7\7");/*到了发出5声蜂鸣,作为提示*/
                }

        }
}
void init_daemon(void)/*这是守护进程*/
{
    pid_t child1,child2;
    int i;
    child1=fork();
    if(child1>0)
            exit(0);
    else
            if(child1< 0)
            {
                    perror("创建子进程失败");
                    exit(1);
            }
            setsid();
            umask(0);
            for(i=0;i< NOFILE;++i)
                    close(i);
            return;
}
```

任务7 *程序调试。*

设计三个并发的守护进程在后台运行,其中第一子进程写守护进程的运行日志记录,第二子进程child2则监控进程中是否有gedit工具调用,第二子进程child3则检查自己是否有新邮件到达，若有则将邮件内容输出到一个主目录下文件中。
```c
#include <stdio.h>
#include <stdlib.h>
#include <sys/types.h>
#include <sys/stat.h>
#include <sys/wait.h>
#include <unistd.h>
#include <syslog.h>
#include <signal.h>
#include <sys/param.h>
#include <time.h>
#include <dirent.h>
  int main()
  {
  pid_t child1,child2,child3;
  struct stat buf;
  int i,check=0,j=0;
  time_t t;
  DIR * dir;
  struct dirent * ptr;
  child1=fork();
  if (child1>0)
  exit(0); /*父进程退出*/
  else if(child1<0)
  {
  perror("创建子进程失败");
  exit(1);
  }
      /*第一子进程*/
  setsid();
  chdir("/");
  umask(0);
  for(i=0;i<NOFILE;++i)
  close(i);
  openlog("守护进程程序信息",LOG_PID,LOG_DAEMON);
  child2=fork();
  if (child2==-1)
  {
  perror("创建子进程失败");
```

```c
exit(2);
}
else if (child2==0)/*第二子进程中的child2*/
{
i=0;
while(i++<100){
system("ps -ef|grep gedit> /home/king/gedit.log");
stat("home/king/gedit.log",&buf);
    if (buf.st_size>180 && check==0)
    {
t=time(0);
syslog(LOG_INFO,"gedit 开始时间为:  %s\n",asctime(localtime(&t)));
check=1;
}
if (buf.st_size<180 && buf.st_size>0 && check==1)
{
t=time(0);
syslog(LOG_INFO,"gedit 结束时间为:  %s\n",asctime(localtime(&t)));
check=0;
}
sleep(1);
}
}
else
{ /*在第一子进程下继续创建进程*/
child3=fork();
if (child3<0){
perror("创建子程序失败");
exit(3);
}
else if (child3==0){ /*第二子进程child3用来查看邮件*/
j=0;
dir=opendir("/var/spool/mail/king");
while (j<6){
j++;
sleep(10);
if ((ptr=readdir(dir))!=NULL){
system("cat /var/spool/mail/king/* > /home/king/mail.log");
}
```

```
    }
    closedir(dir);
}
else
{ /*第一子进程写日志来记录守护进程的运行*/
    t=time(0);
    syslog(LOG_INFO,"守护进程开始时间为： %s\n",asctime(localtime(&t)));
    waitpid(child2,NULL,0);
    waitpid(child3,NULL,0);
    t=time(0);
    syslog(LOG_INFO,"守护进程结束时间为： %s\n",asctime(localtime(&t)));
    closelog();
    while (1)
    sleep(10);
}
}
}
```

任务 8 *程序设计。*

设计一个简单的文件监视器。该程序为守护进程，能够监视/etc/ filemonitor.conf 内列出的文件以及文件夹的修改时间。当检测到某个文件的修改时间发生变化时，程序在系统日志中输出该文件的修改时间。

提示：

```
while (1)
{
  fscanf(fp, "%s", str);
  if (feof(fp)) break;
  strcpy(list[count++], str);
}
fclose(fp);
for (i = 0; i < count; i++)
{
  stat(list[i], &buf);
  mtime[i] = buf.st_mtime;
}
while (1)
{
  for (i = 0; i < count; i++)
```

```
    {
     stat(list[i], &buf);
     if (mtime[i] != buf.st_mtime)
     {
       mtime[i] = buf.st_mtime;
       syslog(LOG_INFO,"%swas modified at %s",list[i], asctime(gmtime(&buf.st_mtime)));
     }
    }
    sleep(5);
  }
}
```

五、实验结果记录

六、实验结果分析

七、实验心得

Linux 程序设计实验报告 9

——进程通信 1

一、实验目的

1. 掌握常用的几种硬中断方法；
2. 了解常用的软中断；
3. 掌握 signal 函数实现信号处理程序；
4. 掌握多信号时的信号处理程序编写；
5. 掌握应用管道实现信号处理的方法。

二、实验任务与要求

1. Ctrl+C 硬中断；
2. alarm 函数产生的 SIGALRM 信号；
3. 应用 signal 函数实现信号处理程序编写；
4. 多信号时的信号处理程序编写；
5. 应用管道实现信号处理的编写；
6. 调用系统函数 kill 对进程的处理。

三、实验工具与准备

PC 机、Linux Redhat Fedora Core6 操作系统。

四、实验步骤与操作指导

任务 1 程序调试。

硬中断实例，程序的功能是通过硬中断的方法中止进程的运行。
运行下列程序 kk.c：

```
#include <unistd.h>
int main(void)
{
```

```
        while(1);
        return 0;
}
```

（1）程序运行过程中，请使用硬中断 Ctrl+C 或 Ctrl-\ 中断程序的执行。

（2）可以使用信号 SIGSEGV 中断此程序，方法是先在后台运行此程序，得出程序进程号，然后用命令 kill 发送信号 SIGSEGV，如下形式：

```
$ ./kk  &  [1] 7940
$ kill -SIGSEGV 7940
$ （再次回车）
```

任务2 *程序调试。*

使用软件中断 alarm 函数和 SIGALRM 信号,调用 alarm 函数设定一个闹钟,告诉内核在 n 秒之后给当前进程发 SIGALRM 信号，该信号的默认处理动作是终止当前进程。这个函数的返回值是 0 或者是以前设定的闹钟时间还余下的秒数。如果 n 值为 0，表示取消以前设定的闹钟，函数的返回值仍然是以前设定的闹钟时间还余下的秒数。

程序源代码 kkk.c:

```
#include <unistd.h>
#include <stdio.h>
int main(void)
{
    int counter;
    alarm(10);
    for(counter=0; 1; counter++)
        printf("counter=%d ", counter);
    return 0;
}
```

程序的作用是 10 秒钟之内不停地计数，10 秒钟到了就被 SIGALRM 信号中断。

任务3 *程序设计。*

设计一个程序，要求程序运行后进入无限循环，在无限循环中每 3 秒输出一条语句；当用户按下中断组合键（Ctrl+C）发送信号 SIGINT，此时调用信号处理函数（自定义函数 fun_ctrl_c）。在程序正常结束前，再应用 signal 函数（用参数 SIG_DFL），恢复系统对信号的默认处理方式。

 提示：

在 main 函数中有：
(void) signal(SIGINT,fun_ctrl_c);

（1）修改程序，要求程序运行后进入一个无限循环，当用户按下中断键 Ctrl+Z 时，进入程序的自定义信号处理函数,当用户再次按下中断键 Ctrl+Z 后,结束程序运行。

(2)修改程序,要求程序运行后进入一个无限循环,当用户按下中断键 Ctrl+Z 时,进入程序的自定义信号处理函数,当用户再次按下中断键 Ctrl+Z 后,程序仍能继续运行。

任务 4 程序设计。

程序的功能是程序运行后进入无限循环,当用户按下中断键 Ctrl+Z 时,程序的自定义信号处理函数输出一幅图形,当用户按下中断键 Ctrl+\ 时产生一批随机数,当用户按下中断键 Ctrl+C 时结束程序运行。

任务 5 调试下列程序。

程序中能处理三种不同的信号,其中信号 SIGINT(Ctrl+C 键)和 SIGTSTP(Ctrl+Z 键)是可阻塞的,而信号 SIGQUIT(Ctrl+\ 键)是不可阻塞的,程序源代码如下:

```c
#include<stdio.h>
#include<stdlib.h>
#include<signal.h>
#include<sys/types.h>
#include<unistd.h>
void fun_ctrl_c();
void fun_ctrl_z();
void fun_ctrl_d();
int main()
{                           /*C 程序的主函数,开始入口 */
    int i;
    sigset_t set, pendset;
    struct sigaction action;
    (void) signal(SIGINT, fun_ctrl_c);        /*调用 fun_ctrl_c 函数 */
    (void) signal(SIGTSTP,fun_ctrl_z);
    (void) signal(SIGQUIT, fun_ctrl_d);
    if (sigemptyset(&set) < 0) /*初始化信号集合 */
    perror("初始化信号集合错误");
    if (sigaddset(&set, SIGINT) < 0) /*把 SIGINT 信号加入信号集合 */
    perror("Ctrl+C 加入信号集合错误");
    if (sigaddset(&set, SIGTSTP) < 0)
    perror("Ctrl+Z 加入信号集合错误");
    if (sigprocmask(SIG_BLOCK, &set, NULL) < 0)
        perror("往信号阻塞集增加一个信号集合错误");
    else
        {
    for (i = 0; i < 10; i++) {
```

```c
        printf("Ctrl+C、Ctrl+Z 信号处理处于阻塞状态，能及时处理'Ctrl+\'信号\n ");
            sleep(3);
    }
    }
        if (sigprocmask(SIG_UNBLOCK, &set, NULL) < 0)
/*当前的阻塞集中删除一个信号集合 */
        perror("从信号阻塞集删除一个信号集合错误");
}

void fun_ctrl_c()    /*自定义信号处理函数 */
{
    int n ;
    printf("\t 你按了 Ctrl+C 系统是不是很长时间没理你？\n");
    for(n=0;n<4;n++)
        printf("\t 正在处理 Ctrl+C 信号处理函数 \n");
}


void fun_ctrl_z()    /*自定义信号处理函数 */
{
    int n ;
    printf("\t 你按了 Ctrl+Z 系统是不是很长时间没理你？\n");
    for(n=0;n<6;n++)
        printf("\t 正在处理 Ctrl+Z 信号处理函数 \n");
}


void fun_ctrl_d( )              /*自定义信号处理函数 */
{
    int n;
    printf("\t 你按了'Ctrl+\' 系统及时地处理了此信号处理函数\n");
    for(n=0;n<2;n++)
        printf("\t 正在处理 Ctrl+/ 信号处理函数 \n");
}
```

问题：
（1）调试此程序，画出此程序的中断机制；
（2）修改此程序，把此程序的框架应用到实际当中。

任务6 程序设计。

设计一个程序要求创建一个管道，复制进程，父进程往管道中写入字符串，子进程从管道中读取前输出字符串。提示主程序调用pipe函数创建一个管道，调用fork函数创建进程；父进程中先用 close(pipe_fd[0])关闭 pipe_fd[0]，剩下的 pipe_fd[1]用来把数据写入管道，利用 write 函数写入字符串，然后用 close(pipe_fd[1])关闭 pipe_fd[1]；子进程是用close(pipe_fd[1])关闭pipe_fd[1]，剩下的 pipe_fd[0]用来从管道读取数据，利用 read 函数读取字符串，然后用 close(pipe_fd[0])关闭 pipe_fd[0]。

```c
#include<stdio.h>
#include<stdlib.h>
#include<sys/types.h>
#include<sys/wait.h>
#include<unistd.h>
#include<string.h>
int main ()
{
  pid_t result;
  int r_num;
  int pipe_fd[2];
  char buf_r[100],buf_w[100];
  memset(buf_r,0,sizeof(buf_r));
  if(pipe(pipe_fd)<0)
  {
    printf("创建管道失败");
    return -1;
  }
  result=fork();
  if(result<0)
  {
    perror("创建子进程失败");
    exit(0) ;
  }
  else if (result==0)              /*子进程运行代码段*/
  {
    close(pipe_fd[1]);
    if((r_num=read(pipe_fd[0],buf_r,100))>0)
   printf("子进程从管道读取%d 个字符，读取的字符串是：%s\n",r_num,buf_r);
    close(pipe_fd[0]);
    exit(0);
  }
```

```
      else                    /*父进程运行代码段*/
      {
        close(pipe_fd[0]);
        printf("请从键盘输入写入管道的字符串\n");
        scanf("%s",buf_w);
        if(write(pipe_fd[1],buf_w,strlen(buf_w))!=-1)
        printf("父进程向管道写入:%s\n",buf_w);
        close(pipe_fd[1]);
        waitpid(result,NULL,0);//调用 waitpid, 阻塞父进程,等待子进程退出
        exit(0);
      }
    }
```

问题：

（1）设计一个程序，要求创建一个管道，创建子进程，父进程读取内容，把读出的字符串写入管道，子进程从管道中读取前输出字符串。

（2）设计一个程序，要求创建一个管道 PIPE，应用函数 fork 复制子进程，在父进程中运行命令"ls –l"，把某一文件名的信息写入管道，子进程从管道中读取这些文件信息。

任务7 阅读下列资料，调试比较各程序，给出调试结果。

编写一个 C 程序，完成以下功能。

（1）父进程创建2个子进程 P1、P2；

（2）父进程捕捉从键盘上通过 CTRL+C 键发来的中断信号；

（3）父进程获得中断信号后使用系统调用 kill()向两个子进程分别发终止执行信号 SIGUSR1和 SIGUSR2；

（4）子进程捕捉到各自的信号后分别输出：

child1 is killed by parent!和 child2 is killed by parent!

然后终止执行（无先后次序的要求）；

（5）父进程等待两个子进程终止后输出以下信息，然后终止执行：

parent process is killed!

五、实验结果记录

六、实验结果分析

七、实验心得

Linux 程序设计实验报告 10

——进程通信 2

一、实验目的

1. 掌握无名管道与命名管道进行通信；
2. 掌握消息队列的读写操作；
3. 掌握内存映射的数据共享；
4. 了解 UNIX System V 共享内存通信机制。

二、实验任务与要求

1. 管道读写程序的编写与应用；
2. 应用命名管道实现聊天程序设计；
3. 应用函数 msgget、msgrcv、msgsnd，读、写消息队列中的消息并打印输出；
4. 应用用消息队列实现简单的聊天程序；
5. 父子进程通过内存映射实现数据共享；
6. UNIX System V 共享内存通信机制程序设计。

三、实验工具与准备

PC 机、Linux Redhat Fedora Core6 操作系统。

四、实验步骤与操作指导

任务 1 程序调试。

设计程序复制一个进程，父进程读取一段源程序的内容，删除其中的注释后通过管道发送给子进程，并呈现在屏幕上。程序中用管道实现了文件的传输，另外删除源代码中的注释也是一个挺实用的功能，这里的注释特指/*、*/与它们之间的内容。

```
#include <stdio.h>
#include <stdlib.h>
#include <unistd.h>
```

```c
#include <sys/wait.h>
#include <sys/types.h>
void f1();                    /*让文件指针跳过注释部分*/
void f2(char c);              /*对单双引号中内容的处理*/
FILE *fp;
char buf_s[1000];
int i=0;                      /*这几个变量在不同函数中均需出现，故定义为全局变量*/
int main()
{
    pid_t pid;
    int pipe_fd[2];
    char buf_r[1000],c,d;
    memset(buf_r,0,sizeof(buf_r));           /*初始化清空*/
    if(pipe(pipe_fd)<0)                       /*创建管道*/
    {
        printf("创建管道失败");
        return -1;
    }
    pid=fork();                               /*创建子进程*/
    if(pid<0)                                 /*fork函数的返回小于0表示出错*/
    {
        perror("创建子进程失败");
        exit;
    }
    else if (pid==0)                          /*返回值为0代表子进程*/
    {
        close(pipe_fd[1]);
        read(pipe_fd[0],buf_r,1000);
        printf("子进程从管道读取的文件内容是：\n%s",buf_r);
        close(pipe_fd[0]);
        exit(0);
    }
    else        /*返回值大于0代表父进程*/
    {
        if((fp=fopen("ceshi.c","r"))==NULL)   /*打开待处理的文件*/
            exit(0);
        while((c=fgetc(fp))!=EOF)
        {
            if(c=='/')
```

```c
                    if((d=fgetc(fp))=='*')
                        f1();                    /*出现/*则调用f1()函数*/
                    else
                    {
                        buf_s[i]=c;
                        buf_s[i+1]=d;
                        i+=2;
                    }                            /*普通字符直接存储*/
                else if(c=='\''||c=='\"')
                    f2(c);                       /*调用f2()处理单、双引号间的内容*/
                else
                {
                    buf_s[i]=c;
                    i++;
                }                                /*普通字符直接存储*/
            }
            buf_s[i]=0;          /*前面已经进行过初始化,这一句可有可无*/
            fclose(fp);
            close(pipe_fd[0]);
            write(pipe_fd[1],buf_s,1000);
            printf("父进程向管道写入完毕!\n");
            close(pipe_fd[1]);
            waitpid(result,NULL,0);
            exit(0);
        }
    }
}
void f1()
{
    char c,d;
    c=fgetc(fp);
    d=fgetc(fp);
    while(c!='*'||d!='/')
    {    c=d;
         d=fgetc(fp); }
}
void f2(char c)
{
    char d;
    buf_s[i]=c;
```

```
        i++;                          /*存入一个单、双引号*/
        while((d=fgetc(fp))!=c)
        {
            buf_s[i]=d;
            i++;
            if(d=='\\')               /*出现转义字符\'或\",\后的单、双引号应直接存储*/
            {   buf_s[i]=fgetc(fp);
                i++;
            }
        }
        buf_s[i]=d;                   /*存入另一个单、双引号*/
        i++;
}
```

任务 2 设计程序。

设计两个程序，要求用命名管道实现聊天程序，每次发言后自动在后面增加当前系统时间。增加结束字符，比如最后输入"88"后结束进程。

任务 3 调试下列程序。

设计一个程序，要求用函数 msgget 创建消息队列，从键盘输入的字符串添加到消息队列，然后应用函数 msgrcv 读取队列中的消息并在计算机屏幕上输出。程序先调用 msgget 函数创建、打开消息队列，接着调用 msgsnd 函数，把输入的字符串添加到消息队列中，然后调用 msgrcv 函数，读取消息队列中的消息并打印输出，最后调用 msgctl 函数，删除系统内核中的消息队列。

```c
#include <stdio.h>
#include <string.h>
#include <stdlib.h>
#include <sys/types.h>
#include <sys/msg.h>
#include <sys/ipc.h>
#include <unistd.h>
struct msgmbuf                        /*结构体，定义消息的结构*/
{
    long msg_type;                    /*消息类型*/
    char msg_text[512];               /*消息内容*/
};
int main()
{
int qid;
```

```c
    key_t key;
    int len;
    struct msgmbuf msg;
    if((key=ftok(".",'a'))==-1)      /*调用ftok函数，产生标准的key*/
    {
    perror("产生标准key出错");
    exit(1);
    }
    if((qid=msgget(key,IPC_CREAT|0666))==-1)
     {
    perror("创建消息队列出错");
    exit(1);
    }
    printf("创建、打开的队列号是：%d\n",qid);   /*打印输出队列号*/
    puts("请输入要加入队列的消息: ");
    if((fgets((&msg)->msg_text,512,stdin))==NULL)/*输入的消息存入变量msg_text*/
    {
    puts("没有消息");
    exit(1);
    }
    msg.msg_type=getpid();
    len=strlen(msg.msg_text);
    if((msgsnd(qid,&msg,len,0))<0)    /*调用msgsnd函数，添加消息到消息队列*/
    {
    perror("添加消息出错");
    exit(1);
    }
    if((msgrcv(qid,&msg,512,0,0))<0)   /*调用msgrcv函数，从消息队列读取消息*/
    {
    perror("读取消息出错");
    exit(1);
    }
    printf("读取的消息是：%s\n",(&msg)->msg_text);  /*打印输出消息内容*/
    if((msgctl(qid,IPC_RMID,NULL))<0)/*调用msgctl函数，删除系统中的消息队列*/
    {
    perror("删除消息队列出错");
    exit(1);
    }
    exit (0);
```

}

问题：
（1）修改程序，过滤一些字符串；把读取的信息存入文件中。
（2）是否能够从一个进程写数据，在另一个进程读数据？

任务4 设计程序。
设计两个程序要求用消息队列实现简单的聊天功能。

任务5 调试程序。
在主程序中先调用 mmap 映射内存，然后调用 fork 函数创建进程。在调用 fork 函数之后，子进程继承父进程匿名映射后的地址空间，父子进程通过映射区域实现共享内存。
程序代码如下：

```c
#include<sys/types.h>
#include<unistd.h>
#include <sys/mman.h>
#include <fcntl.h>
typedef struct
{
  char name[4];
  int  age;
}people;
int main(int argc, char** argv)
{
pid_t result;
int i;
people *p_map;
char temp;
p_map=(people*)mmap(NULL,sizeof(people)*10,PROT_READ|PROT_WRITE,
MAP_SHARED|MAP_ANONYMOUS,-1,0);        /*调用mmap函数，匿名内存映射*/
result=fork();            /*调用fork函数，复制进程，返回值存在变量result中*/
if(result<0)  /*通过result的值来判断fork函数的返回情况，这儿进行出错处理*/
{
perror("创建子进程失败");
exit(0);
}
else if (result==0)       /*返回值为0代表子进程*/
{
```

```
sleep(2);
for(i = 0;i<5;i++)
printf("子进程读取：第 %d 个人的年龄是： %d\n",i+1,(*(p_map+i)).age);
(*p_map).age = 110;
munmap(p_map,sizeof(people)*10);   /*解除内存映射关系*/
exit(0);
}
else                        /*返回值大于 0 代表父进程*/
{
temp = 'a';
for(i = 0;i<5;i++)
{
temp += 1;
memcpy((*(p_map+i)).name, &temp,2);
(*(p_map+i)).age=20+i;
}
sleep(5);
printf( "父进程读取：五个人的年龄和是： %d\n",(*p_map).age );
printf("解除内存映射……\n");
munmap(p_map,sizeof(people)*10);
printf("解除内存映射成功！\n");
}
return 0;
}
```

问题：

（1）设计两个程序，要求用 mmap 系统调用，通过映射共享内存，实现简单的字符交换；

（2）修改程序，对接收的字符进行判断，然后决定是否中止进程程序。

任务 6 调试程序。

程序的功能是通过 UNIX System V 共享内存通信机制，进行相互通信的程序。

10-6-1.c 程序代码如下：

```
#include <sys/ipc.h>
#include <sys/shm.h>
#include <sys/types.h>
#include <unistd.h>
#include <stdio.h>
#include <time.h>
```

```c
typedef struct
{
    char str1[50];
    char str2[50];
    int no1,no2;
} str;
main(int argc, char** argv)
{
    int shm_id,i;
    key_t key;
    str *p_map;
    time_t timep;
    struct tm *p;
    char* name = "/dev/shm/myshm2";
    key = ftok(name,0);                    /*调用ftok函数，产生标准的key*/
    shm_id=shmget(key,4096,IPC_CREAT);
    if(shm_id==-1)
    {
     perror("获取共享内存区域的ID出错");
       return 0;
    }
    p_map=(str*)shmat(shm_id,NULL,0);/* 调用shmat函数，映射共享内存*/
    (*p_map).no1=0;(*p_map).no2=0;
    printf("This is A.\n");
    fgets((*p_map).str2,sizeof((*p_map).str2),stdin);
    (*p_map).no1=1;
    while(1)
    {
    if ((*p_map).no2==1)
    {
      time(&timep);
      gmtime(&timep);
      p=localtime(&timep);
     printf("B:%s%d:%d:%d\n",(*p_map).str1,p->tm_hour,p->tm_min,p->tm_sec);
      (*p_map).str1[0]='\0';
       (*p_map).no2=0;
    }
    else continue;
```

```c
    if ((*p_map).no1==0)
      {
       fgets((*p_map).str2,sizeof((*p_map).str2),stdin);
          (*p_map).no1=1;
          }
  }
if(shmdt(p_map)==-1)          /*调用 shmdt 函数，解除进程对共享内存区域的映射*/
    perror("解除映射出错");
}
```

10-6-2.c 程序代码如下：

```c
#include <sys/ipc.h>
#include <sys/shm.h>
#include <sys/types.h>
#include <unistd.h>
#include <stdio.h>
#include <time.h>
typedef struct
{
    char str1[50];
    char str2[50];
    int no1,no2;
} str;
main(int argc, char** argv)
{
    int shm_id,i;
    key_t key;
    str *p_map;
    time_t timep;
    struct tm *p;
    char* name = "/dev/shm/myshm2";
    key = ftok(name,0);  /*调用 ftok 函数，产生标准的 key*/
    shm_id = shmget(key,4096,IPC_CREAT);
    if(shm_id == -1)
    {
            perror("获取共享内存区域的 ID 出错");
            return;
    }
    p_map = (str*)shmat(shm_id,NULL,0);
    (*p_map).no1=0; (*p_map).no2=0;
```

```c
        printf("This is B.\n");
        while(1)
        {
        if ((*p_map).no1==1)
         {
           time(&timep);
          gmtime(&timep);
          p=localtime(&timep);
         printf("A:%s%d:%d:%d\n",(*p_map).str2,p->tm_hour,p->tm_min,p->tm_sec);
          (*p_map).str2[0]='\0';
           (*p_map).no1=0;
         }
        else continue;
        if ((*p_map).no2==0)
        {
           fgets((*p_map).str1,sizeof((*p_map).str1),stdin);
           (*p_map).no2=1;
          }
        }
        if(shmdt(p_map) == -1)  /*调用 shmdt 函数，解除进程对共享内存区域的映射*/
         perror("解除映射出错");
        }
```

问题：

（1）调试此程序，画出此程序的通信机制；

（2）修改此程序，把此程序的框架应用到实际当中。要求设计两个程序，用系统 V 共享内存通信，实现简单的聊天功能，具体要求如下：

1）为两个程序建立系统 V 共享内存；

2）每个程序用两个进程分别控制输入输出，达到即时通讯的效果；

3）支持显示系统时间，支持输入 QUIT 退出程序。

五、实验结果记录

六、实验结果分析

七、实验心得

Linux 程序设计实验报告 11

——Linux 线程程序设计

一、实验目的

1. 了解 Linux 下线程和进程的概念；
2. 了解多线程程序的基本原理；
3. 了解 pthread 库；
4. 掌握临界区的概念及临界区的处理；
5. 学会使用 pthread 库中的函数编写多线程程序。

二、实验任务与要求

1. 新线程的创建，以及和主线程之间的关系；
2. 线程之间的通信和共享变量；
3. 临界区的处理；
4. 编写一个生产者–消费者程序；
5. 用多种方法实现线程之间的同步。

三、实验工具与准备

PC 机、Linux Redhat Fedora Core6 操作系统。

四、实验步骤与操作指导

任务 1 调试下列程序。

程序中使用 pthread 线程库创建一个新线程，在父进程（也可以称为主线程）和新线程中分别显示进程 id 和线程 id，并观察线程 id 数值。

程序代码如下：

```
#include <stdio.h>
#include <string.h>
#include <stdlib.h>
```

```c
#include <pthread.h>    /*pthread_create()函数的头文件*/
#include <unistd.h>

pthread_t ntid;
void printids(const char *s)   /*各线程共享的函数*/
{
    pid_t    pid;
    pthread_t  tid;
    pid = getpid();
    tid = pthread_self();
    printf("%s  pid=  %u  tid= %u  (0x%x)\n", s, (unsigned int)pid,
(unsigned int)tid, (unsigned int)tid);
}

void *thread_fun(void *arg)  /*新线程执行代码*/
{
    printids(arg);
    return NULL;
}

int main(void)
{
    int err;
/*下列函数创建线程*/
    err = pthread_create(&ntid, NULL, thread_fun, "我是新线程: ");
    if (err != 0) {
            fprintf(stderr, "创建线程失败: %s\n", strerror(err));
            exit(1);
    }
    printids("我是父进程:");
    sleep(2);/*挂起2秒，等待新线程运行结束*/
    return 0;
}
```

问题：

（1）进程在一个全局变量 ntid 中保存了新创建的线程的 id，如果新创建的线程不调用 pthread_self 而是直接打印这个 ntid，能不能达到同样的效果？

（2）在本题中，如果没有 sleep(5)函数，会出现什么结果？

（3）运行以下程序，能得出什么结论？

```
#include<stdio.h>
#include<pthread.h>
void *print_thread_id(void *arg)
{
    /* 打印当前线程的线程号*/
    printf("Current thread id is %u\n", (unsigned)pthread_self());
}

int main(int argc, char *argv[])
{
    pthread_t thread;                    /*保存线程号*/
    pthread_create(&thread, NULL, print_thread_id, NULL);    /*创建一个线程*/

    sleep(10);        /*休眠1s*/
    /*打印进程号    */
    printf("Main thread id is %u\n", (unsigned)pthread_self());
    return 0;
}
```

任务 2 应用 pthread_create 函数，启动一个线程，在线程中执行一个循环输出，在主进程中执行一个较长时间执行的任务。

任务 3 调试下列程序。程序只是给出线程产生、退出和等待的程序。

程序代码如下：

```
#include <stdio.h>
#include <string.h>
#include <stdlib.h>
#include <pthread.h>    /*pthread_create 函数的头文件*/
#include <unistd.h>
void * thread_fun1(void *arg)
{
    printf("thread 1 returning\n");
    return((void *)1);
```

```
}

void * thread_fun2(void *arg)
{
    printf("thread 2 exiting\n");
    pthread_exit((void *)2);
}

int main(void)
{
    int        err;
    pthread_t  tid1, tid2;
    void       *tret;
    err = pthread_create(&tid1, NULL, thread_fun1, NULL);  /*创建第1个线程*/
    if (err != 0)
        fprintf(stderr,"can't create thread 1: %s\n", strerror(err));
    err = pthread_create(&tid2, NULL, thread_fun2, NULL);  /*创建第2个线程*/
    if (err != 0)
        fprintf(stderr,"can't create thread 2: %s\n", strerror(err));
    err = pthread_join(tid1, &tret);  /*等待第1个线程结束*/
    if (err != 0)
        fprintf(stderr,"can't join with thread 1: %s\n", strerror(err));
    printf("thread 1 exit code %d\n", (int)tret);
    err = pthread_join(tid2, &tret); /*等待第2个线程结束*/
    if (err != 0)
        fprintf(stderr,"can't join with thread 2: %s\n", strerror(err));
    printf("thread 2 exit code %d\n", (int)tret);
    exit(0);  /* 现在可以安全地返回了 */
}
```

问题：

（1）修改程序，在线程1、2中完成一些任务，例如各自做1到1000的加法，有明显的时间跨度，观察线程的执行情况与线程结束时的状态；

（2）改变线程1、2执行任务的时间，观察结果。

任务4 程序设计。

编写一个多线程程序来生成Fibonacci序列。

任务5 调试下列程序。

```c
#include <stdio.h>
#include <stdlib.h>
#include <unistd.h>
#include <pthread.h>

pthread_mutex_t mutex;  /*定义互斥锁*/
int lock_var; /*两个线程都能修改的共享变量,访问改变量必须互斥*/
void pthread1(void *arg);
void pthread2(void *arg);
int main(int argc, char *argv[])
{
    pthread_t id1,id2;
    int ret;
    pthread_mutex_init(&mutex,NULL); /*互斥锁初始化*/
    ret=pthread_create(&id1,NULL,(void *)pthread1, NULL); /*创建第1个线程*/
    if(ret!=0)
            printf("pthread cread1\n");
    ret=pthread_create(&id2,NULL,(void *)pthread2, NULL); /*创建第2个线程*/
    if(ret!=0)
            printf ("pthread cread2\n");
    pthread_join(id1,NULL); /*等待第1个线程结束*/
    pthread_join(id2,NULL); /*等待第2个线程结束*/
    pthread_mutex_destroy(&mutex); /*释放mutex资源*/
    exit(0);
}

void pthread1(void *arg) /*第1个线程执行代码*/
{
    int i;
    for(i=0;i<2;i++){
        pthread_mutex_lock(&mutex);  /*锁定临界区*/
        /*临界区*/
        lock_var++;
printf("pthread1:第%d次循环,第1次打印 lock_var=%d\n",i,lock_var);
        sleep(1);
printf("pthread1:第%d次循环,第2次打印 lock_var=%d\n",i,lock_var);
    /* 已经完成了临界区的处理,解除对临界区的锁定。*/
        pthread_mutex_unlock(&mutex);   /*解锁*/
```

```
        sleep(1);
      }
   }

   void pthread2(void *arg)   /*第2个线程执行代码*/
   {
      int i;
      for(i=0;i<5;i++){
         pthread_mutex_lock(&mutex);   /*锁定临界区*/
          /*临界区*/
         sleep(1);
         lock_var++;
     printf("pthread2：第%d 次循环 lock_var=%d\n",i,lock_var);
          /* 已经完成了临界区的处理，解除对临界区的锁定。*/
         pthread_mutex_unlock(&mutex);   /*解锁 */
         sleep(1);
      }
   }
```

 问题：

（1）把程序 pthread1、pthread2 中所有的 pthread_mutex_lock (&mutex) 和 pthread_mutex_unlock(&mutex)代码除去，也就是不对访问共享变量 lock_var 进行互斥。编译并运行程序，观察 pthread1 同一次循环的第 1 次打印和第 2 次打印的结果是否相同？如果在应用程序中产生这样的情况，会发生什么后果？

（2）修改程序，如果在程序中省略语句 sleep(2);对程序运行结果有什么影响？

任务6 调试程序。

程序的功能演示了一个生产者-消费者的例子，生产者生产的产品存放在链表的表头上，消费者从表头取走产品。

```
#include <stdlib.h>
#include <pthread.h>
#include <stdio.h>

struct msg {
   struct msg *next;
   int num;
};
```

```c
struct msg *head;
pthread_cond_t has_product = PTHREAD_COND_INITIALIZER;  /*条件变量置初值*/
pthread_mutex_t lock = PTHREAD_MUTEX_INITIALIZER; /*互斥锁置初值*/

void *producer(void *p)  //生产者线程代码
{
    struct msg *mp;
    int i;
    for (i=0;i<20;++i) {
        mp = malloc(sizeof(struct msg));
        mp->num = rand() % 1000 + 1;
        printf("Produce %d\n", mp->num);
        pthread_mutex_lock(&lock);
        mp->next = head;
        head = mp;
        pthread_mutex_unlock(&lock);
        pthread_cond_signal(&has_product); //唤醒消费者线程
        sleep(rand() % 5);
    }
}

void *consumer(void *p)  //消费者线程代码
{
    struct msg *mp;
    int i;
    for (i=0;i<20;++i)
    {
        pthread_mutex_lock(&lock);
        while (head == NULL)
            pthread_cond_wait(&has_product, &lock);
        mp = head;
        head = mp->next;
        pthread_mutex_unlock(&lock);
        printf("Consume %d\n", mp->num);
        free(mp);
        sleep(rand() % 5);
    }
}
```

```c
int main(int argc, char *argv[])
{
    pthread_t pt, ct;

    srand(time(NULL));
    pthread_create(&pt, NULL, producer, NULL);
    pthread_create(&ct, NULL, consumer, NULL);
    pthread_join(pt, NULL);
    pthread_join(ct, NULL);
    return 0;
}
```

问题：

（1）调试此程序，画出此程序的通信机制；

（2）修改此程序，把此程序的框架应用到实际当中。

任务7 程序设计。

有两个线程T1和T2动作描述如下，x、y、z为两个线程共享变量。信号量S1和S2的初值均为0，编写程序完成下面两个线程T1、T2并发执行。不断调整sleep(n)的n值，多次运行程序后，观察x、y、z输出的值，各为多少？

五、实验结果记录

六、实验结果分析

七、实验心得

Linux 程序设计实验报告 12

——Linux 网络程序设计

一、实验目的

1. 掌握端口及 socket 的使用；
2. 掌握面向连接的 TCP 编程；
3. 掌握面向非连接的 UDP 编程；
4. 掌握 I/O 多路利用的控制；
5. 掌握复杂网络程序的实现。

二、实验任务与要求

1. 网络参数的获取；
2. TCP 协议下网络通信程序设计；
3. 客户端与服务器端网络通信程序设计；
4. 文件传送的网络程序设计；
5. UDP 方式的网络程序设计。

三、实验工具与准备

PC 机、Linux Redhat Fedora Core6 操作系统。

四、实验步骤与操作指导

任务 1 调试下列程序。

```
#include <stdio.h>
#include <stdlib.h>
#include <string.h>
#include <unistd.h>
#include <sys/socket.h>
#include <netdb.h>
```

```c
int main(int argc , char *argv[])
{
  struct hostent *host;
  char hostname[]="www.163.com";
  struct in_addr in;
  struct sockaddr_in  addr_in;
  extern int h_errno;
  if((host=gethostbyname(hostname))!=NULL)
  {
   memcpy(&addr_in.sin_addr.s_addr,host->h_addr,4);
   in.s_addr=addr_in.sin_addr.s_addr;
   printf("Domain  name : %s \n",hostname);
   printf("IP    length : %d\n" ,host->h_length);
   printf("Type         : %d\n" ,host->h_addrtype);
   printf("IP           : %s\n" ,inet_ntoa(in));
  }
  else
  {
   printf("Domain  name : %s \n",hostname);
   printf("error        : %d\n" ,h_error);
   printf("             : %s\n" ,hstrerror(h_error));
  }
}
```

问题：

（1）编译程序，程序中有两个错误，请改正。

（2）运行程序，在网络与Linux操作系统断开与连接2种情况下，程序的运行结果分别是什么？

（3）程序中显示的IP   length 与 Type 及外部变量 h_errno 分别表示什么含义？

（4）修改程序，注释掉 char hostname[]="www.163.com"; 这一行，把程序中所需的域名在程序运行时终端以命令行参数形式输入。

任务2 程序设计。

编写一个基于TCP协议的网络通信程序，要求服务器通过socket连接后，并要求输入用户，判断为 liu 时，才向客户端发送字符串"Hello, you are connected!"。在服务器上显示客户端的 IP 地址或域名。

任务3 调试下列程序。

server.c 的作用是从客户端读字符，然后将每个字符转换为大写并回送给客户端。

```c
/* server.c */
#include <stdio.h>
#include <stdlib.h>
#include <string.h>
#include <unistd.h>
#include <sys/socket.h>
#include <netinet/in.h>

#define MAXLINE 80
#define SERV_PORT 8000

int main(void)
{
    struct sockaddr_in servaddr, cliaddr;
    socklen_t cliaddr_len;
    int listenfd, connfd;
    char buf[MAXLINE];
    char str[INET_ADDRSTRLEN];
    int i, n;

    listenfd = socket(AF_INET, SOCK_STREAM, 0);
    bzero(&servaddr, sizeof(servaddr));
    servaddr.sin_family = AF_INET;
    servaddr.sin_addr.s_addr = htonl(INADDR_ANY);
    servaddr.sin_port = htons(SERV_PORT);
    bind(listenfd, (struct sockaddr *)&servaddr, sizeof(servaddr));
    listen(listenfd, 20);
    printf("Accepting connections ...\n");
    while (1)
    {
        cliaddr_len = sizeof(cliaddr);
        connfd = accept(listenfd,
            (struct sockaddr *)&cliaddr, &cliaddr_len);
        n = read(connfd, buf, MAXLINE);
        printf("received from %s at PORT %d\n",
            inet_ntop(AF_INET, &cliaddr.sin_addr, str, sizeof(str)),
            ntohs(cliaddr.sin_port));
```

```c
        for (i = 0; i < n; i++)
            buf[i] = toupper(buf[i]);
        write(connfd, buf, n);
        close(connfd);
    }
}
```

client.c 的作用是从命令行参数中获得一个字符串发给服务器,然后接收服务器返回的字符串并打印。

```c
/* client.c */
#include <stdio.h>
#include <stdlib.h>
#include <string.h>
#include <unistd.h>
#include <sys/socket.h>
#include <netinet/in.h>

#define MAXLINE 80
#define SERV_PORT 8000

int main(int argc, char *argv[])
{
    struct sockaddr_in servaddr;
    char buf[MAXLINE];
    int sockfd, n;
    char *str;

    if (argc != 2)
    {
            fputs("usage: ./client message\n", stderr);
            exit(1);
    }
    str = argv[1];
    sockfd = socket(AF_INET, SOCK_STREAM, 0);
    bzero(&servaddr, sizeof(servaddr));
    servaddr.sin_family = AF_INET;
    inet_pton(AF_INET, "127.0.0.1", &servaddr.sin_addr);
    servaddr.sin_port = htons(SERV_PORT);
    connect(sockfd, (struct sockaddr *)&servaddr, sizeof(servaddr));
```

```
        write(sockfd, str, strlen(str));
        n = read(sockfd, buf, MAXLINE);
        printf("Response from server:\n");
        write(STDOUT_FILENO, buf, n);
        close(sockfd);
        return 0;
}
```

问题：

（1）编辑、编译程序，用以下方式执行程序：

a.在服务器运行：

\#./server

　　\# netstat –apn|grep 8000

b.在客户端运行：

　　\#./client abcd

回到 server 所在的终端，看看 server 的输出，再次运行服务器端程序：

　　\#./server

客户端的端口号是自动分配的。现在把客户端所连接的服务器 IP 改为其它主机的 IP，试试两台主机的通讯。将 Client 端的地址改为外网地址后，运行结果如何？

（2）在客户端的 connect()代码之后插一个 while(1);死循环，使客户端和服务器都处于连接中的状态，用 netstat 命令查看。

a.在服务器运行：

 \# ./server &

　　\# Accepting connections …

b.在客户端运行：

　　\#./client abcd &

　　\# netstat –apn|grep 8000

任务 4 调试下列程序。

```
#include<stdio.h>
#include<stdlib.h>
#include<string.h>
#include<sys/socket.h>
#include<netinet/in.h>
#include<arpa/inet.h>
#include<netdb.h>
#include<errno.h>
#include<sys/types.h>
```

```c
int port=8888;
int main()
{
    int sockfd;
    int i=0;
    int z;
    char buf[80],str1[80];
    struct sockaddr_in adr_srvr;
    FILE *fp;
    printf("打开文件......\n");
    /*以只读的方式打开liu文件*/
    fp=fopen("liu","r");
    if(fp==NULL)
    {
        perror("打开文件失败");
        exit(1);
    }
    printf("连接服务端...\n");
    /* 建立IP地址 */
    adr_srvr.sin_family=AF_INET;
    adr_srvr.sin_port=htons(port);
    adr_srvr.sin_addr.s_addr = htonl(INADDR_ANY);
    bzero(&(adr_srvr.sin_zero),8);
    sockfd=socket(AF_INET,SOCK_DGRAM,0);
    if(sockfd==-1)
    {
        perror("socket 出错");
        exit(1);
    }
    printf("发送文件 ....\n");
    /* 读取三行数据,传给udpserver*/
    for(i=0;i<3;i++)
    {
        fgets(str1,80,fp);
        printf("%d:%s",i,str1);
        sprintf(buf,"%d:%s",i,str1);
        z=sendto(sockfd,buf,sizeof(buf),0,(struct sockaddr *)&adr_srvr,sizeof(adr_srvr));
        if(z<0)
```

```
        {
        perror("recvfrom 出错");
        exit(1);
        }
     }
     printf("发送.....\n");
     sprintf(buf,"stop\n");
     z=sendto(sockfd,buf,sizeof(buf),0,(struct sockaddr *)&adr_srvr,
     sizeof(adr_srvr));
     if(z<0)
       {
        perror("sendto 出错");
        exit(1);
       }
     fclose(fp);
     close(sockfd);
     exit(0);
    }
```

任务 5 程序设计。

编写一个以客户机/服务器模式工作的程序,要求在客户端读取系统文件/etc/passwd内容,传送到服务端,服务器端接收字符串,并在显示器显示出来。

任务 6 程序设计题。

编写一个用于两人之间的通讯程序:双方都可以从终端上输入一个字符串,程序将时间信息加入该字符串,然后通过 UDP 的方式发送到对方。程序在运行用须输入本地、目标的 IP 地址和端口号。

五、实验结果记录

六、实验结果分析

七、实验心得

Linux 程序设计实验报告 13

——Linux 图形程序设计

一、实验目的

1. 掌握 SDL 图形开发库的安装；
2. 掌握图形程序设计中图形模式的初始化及基本绘图函数的应用；
3. 掌握图片的装载与显示、图形模式下的文字显示；
4. 初步掌握动画显示的原理；
5. 了解三维绘图的程序设计；
6. 初步掌握 SDL 游戏程序的设计。

二、实验内容

1. SDL 库的安装；
2. 初始化图形模式及背景色的基本设置；
3. 数学图形绘制及运动图形程序编写；
4. 图像的加载与显示的程序设计；
5. 图片显示；
6. 图形、图像、文字程序的设计；
7. 游戏程序设计初步；
8. 动画程序设计初步；
9. 使用 curses.h 库函数图形设计程序初步。

三、实验设备

PC 机、Linux 操作系统。

四、实验步骤与操作指导

任务 1 Linux 操作系统下图形环境的安装。

1. SDL 库的安装

（1）获取 Red Hat 9.0 的安装光盘或是 iso 镜像文件。
（2）点击主菜单|[系统设置][添加/删除应用程序]，如图 1 所示。
（3）找到 x 软件开发，如图 2 所示。

图 1

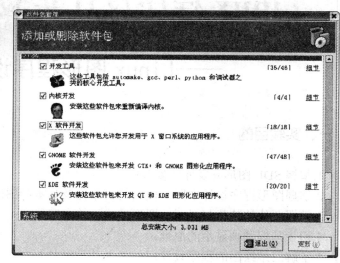

图 2

（3）单击[细节]链接，选中 SDL 相关的选项，如附图 3 所示，在/usr/include/SDL 下可以找到相关的头文件，如附图 4 所示。

2.SDL_draw 库的安装

（1）从 http://sourceforge.net/projects/sdl-draw/ 上下载 SDL_draw 库文件，本实验使用的版本是 SDL_draw-1.2.11，存放在/home/cx/目录下。

（2）解压文件。

```
[root@localhost cx]# tar -vxzf SDL_draw-1.2.11.tar.gz
```

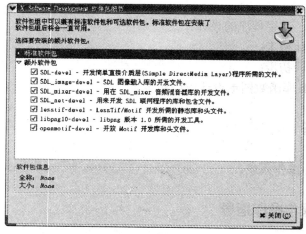

图 3

图 4

解压之后，文件夹中包含的文件如图 5 所示。

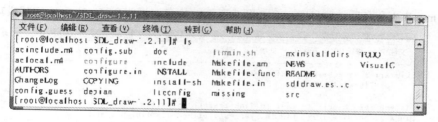

图 5

（3）进入 SDL_draw-1.2.11 文件夹，运行 configure 文件。

```
[root@localhost cx]#cd SDL_draw-1.2.11
[root@localhost SDL_draw-1.2.11]#./configure
```

运行 configure 文件之后，会增加如 Makefile 等文件，可与图 5 比较，如图 6 所示。

图 6

（4）编辑文件 make。

```
[root@localhost SDL_draw-1.2.11]#make
```

如图 7 所示。

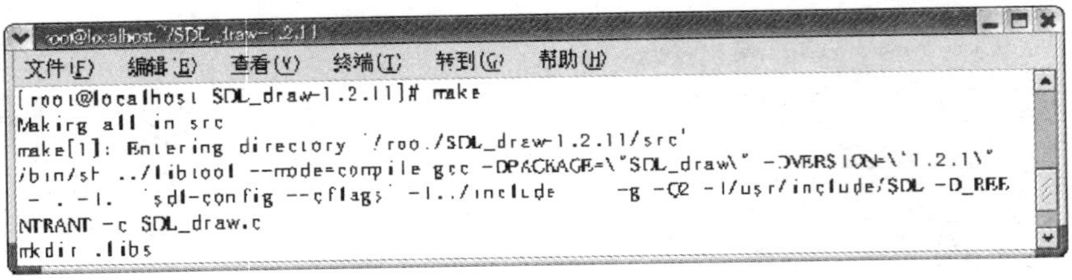

图 7

（5）make install 安装。

`[root@localhost SDL_draw-1.2.11]#make install`

如图 8 所示。

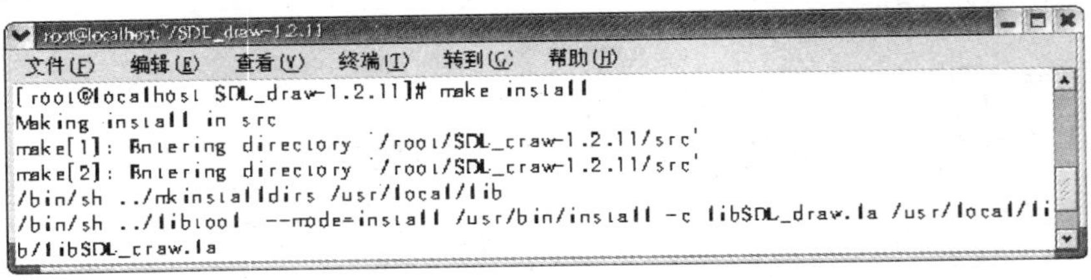

图 8

make install 命令之后，信息中包含如图 9 所示的信息，说明 SDL_draw 库安装成功。

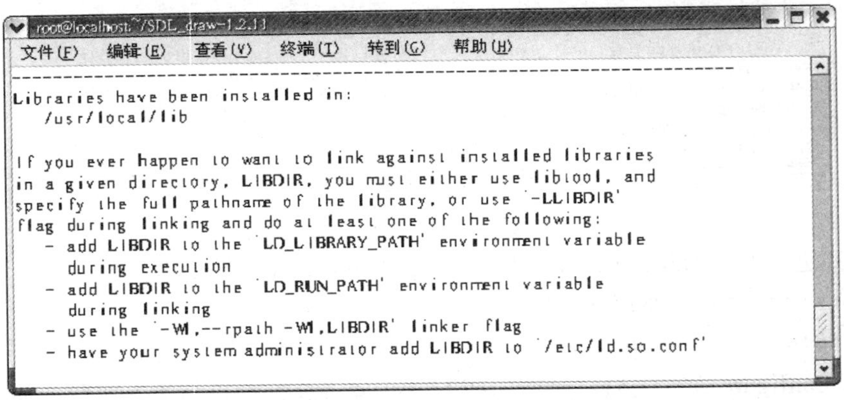

图 9

3.SDL_ttf 的安装

SDL_ttf 是一个支持 TrueType 字体的附加库，从 http://www.libsdl.org/projects/SDL_ttf/ 上可以下载到 SDL_ttf-2.0.8-1.i386.rpm 软件包，用作开发的话还需要下载其开发包 SDL_ttf-devel-2.0.8-1.i386.rpm，存放在/home/cx/目录下。

（1）rpm 安装软件包。

`[root@localhost cx]# rpm -ivh SDL_ttf-2.0.8-1.i386.rpm`

出现如图 10 所示的信息，表示安装成功。

图 10

（2）rpm 安装开发包。

`[root@localhost cx]# rpm -ivh SDL_ttf-devel-2.0.8-1.i386.rpm`

出现如图 11 所示的信息，表示安装成功。

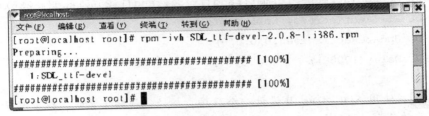

图 11

安装完成之后，可以到 /usr/include/SDL 中查看，如图 12 所示，可以找到对应的 SDL_ttf.h 文件。

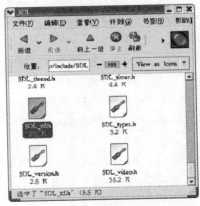

图 12

任务2　初始化视频子系统，设置显示模式为 640×480 大小。设置初始颜色为红色并对颜色值进行改变，渐渐变为白色，停留 5 秒后渐渐变为绿色，最后要求背景色的红、绿、兰为随机显示值，让屏幕停留 10 秒，请编写程序。

提示：

可参考以下关键代码：

```
for(x = 255; x>=0; x--) {
    SDL_MapRGB(screen->format,x,0,0);
    SDL_FillRect(screen, NULL, color);
```

```
            SDL_UpdateRect(screen,0,0,0,0);
        }
        SDL_Delay(5000);
        for(x = 0; x < 256; x++) {
            SDL_MapRGB(screen->format,0,x,0);
            SDL_FillRect(screen, NULL, color);
            SDL_UpdateRect(screen,0,0,0,0);
        }
        SDL_MapRGB(screen->format,rand()%256,rand()%256,rand()%256);
        SDL_FillRect(screen, NULL, color);
        SDL_UpdateRect(screen,0,0,0,0);
        SDL_Delay(10000);
```

任务 3 使用 SDL_draw 库设计一个程序，初始化视频子系统，设置显示模式为 640×480,画面的色深为 16 位，画一个圆，此圆沿着正弦曲线运动。

提示：

请参考下列代码：

```
for(i=0;i<120;i++,x=x+3,y=y+sin(x-80))
{
  Draw_Circle(screen,x,y, r, SDL_MapRGB(screen->format, 255,0,0));
  SDL_Delay(100);
  Draw_Circle(screen,x,y, r, SDL_MapRGB(screen->format, 255,255,255));
  Draw_Line(screen,   x,y,   x+3,y=y+sin(x-80),SDL_MapRGB(screen->format, 255,0,0));
    SDL_UpdateRect(screen, 0, 0, 0, 0);       /*更新整个屏幕*/
}
```

任务 4 设计一个程序，初始化视频子系统，设置显示模式为 640×480，画面的色深为 16 位，加载位图 b.bmp，位图与文件同时存放在/home/cx/下。在程序中编写一个 showBMP()函数，用来显示位图，在主函数中调用它。showBMP()函数中给位图分配一个 surface，得到位图的大小。在主函数中先初始化视频系统，设置视频模式，指定第一张图片的位置。

```
#include<SDL.h>
#include<stdlib.h>
void ShowBMP(char *pn,SDL_Surface * screen,int x,int y)
{
  SDL_Surface *image;
  SDL_Rect dest;
```

```
  image=SDL_LoadBMP(pn);
  if(image==NULL)
  {
    fprintf(stderr,"无法加载 %s:%s\n",pn,SDL_GetError());
    return;
  }
  dest.x=x;
  dest.y=y;
  dest.w=image->w;
  dest.h=image->h;
  SDL_BlitSurface(image,NULL,screen,&dest);
  SDL_UpdateRects(screen,1,&dest);
}

int main()
{
  SDL_Surface *screen;
  int x,y;
  if(SDL_Init(SDL_INIT_VIDEO)<0){
    fprintf(stderr,"无法初始化 SDL: %s\n",SDL_GetError());
    exit(1);
  }
  screen=SDL_SetVideoMode(640,480,16,SDL_SWSURFACE);
  if(screen==NULL){
    fprintf(stderr,"无法设置 640X480X16 位色的视频模式 %s\n",SDL_GetError());
  }
  atexit(SDL_Quit);
  ShowBMP("b.bmp",screen,50,50);
  SDL_Delay(5000);
  return 0;
}
```

任务5 程序设计。

对按一定规律命名的一系列图片，使这些图片自动显示。提示：建立一个包含文件名的数组，用 for 循环依次读入并显示，把文件名和后缀名分开处理。例如：

```
char file[5][6] = {"a","b","c","d","e"} , sub[5] = ".bmp";
int i;
for(i = 0; i < 5; i++){
strcat(file[i],sub);
```

```
        ShowBMP(file[i],screen,50,50);
        SDL_Delay(5000);        /*让屏幕停留 5 秒钟的时间*/
    }
```

任务6　编写一个程序。

运用绘图、位图与文字显示的知识，实现看图识字的效果。即画一个圆，用文字标志"圆"，用位图形式导入一些图片，例如苹果位图，文字标志"苹果"，程序运行结果如下图所示。注意：具体图形可根据具体情况变化。

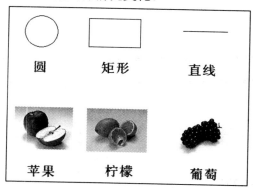

任务7　调试下列程序。

调用了 Linux 系统函数，类似于扫雷程序，程序中要判断某一个坐标是否有事件发生，用闪烁表示。

```c
#include<stdio.h>
#include<stdlib.h>
#include<math.h>
struct Pos
{
    int x;
    int y;
};

main()
{
int i = 0,j = 0,k = 0;
int rand_t = 0,rand_x = 0,rand_y = 0;
int n = 0;
int color = 0;
struct Pos *cursor = NULL;
int *time = NULL;
system("clear");
```

```
printf("\033[49;30minput number of points : \033[?25h");
scanf("%d",&n);
if(n <= 0)
return;    //n<0 就直接结束程序
cursor = (struct Pos*)malloc(n * sizeof(struct Pos));
time = (int *)malloc(n * sizeof(int));
for(i = 0;i < n;i++)
{
while(1)
{
rand_x = abs(rand()) % n + 1;
rand_y = abs(rand()) % n + 1;
//产生两个随机数，准备分别作为点 i 的 x,y 坐标．
for(k = 0;k < i;k++)
{
if(cursor[k].x == rand_x && cursor[k].y == rand_y)
break;
}
    //检验产生的一对坐标与之前的点是否有重。若有，跳出循环。此时 k<i．
if(k == i)  //若 k=i，说明新坐标没有与已有的任何点重合。故可以赋值给新的点。
break;
}
while(1)
{
rand_t = abs(rand()) % (2 * n) + 1;
for(k = 0;k < i;k++)
{
if(time[k] == rand_t)
break;
}
if(k == i)
break;
}
//同上，令时间节点不同。
j = i;
while(j - 1 >= 0 && time[j - 1] > rand_t)
{
time[j] = time[j - 1];
cursor[j] = cursor[j - 1];
```

```
			j--;
		}
		//将各个时间节点排序
		time[j] = rand_t;
		cursor[j].x = rand_x;
		cursor[j].y = rand_y;
	}
	system("clear");
	printf("\033[?25l");
	for(i = 0;i < n;i++)
	{
		color = 30 + abs(rand()) % 9;  //随机产生颜色
		printf("\033[%dm\033[%d;%dH*",color,cursor[i].x,cursor[i].y);
	}
	//产生点

	setbuf(stdout,NULL);
	sleep(time[0]);
	printf("\033[%d;%dH ",cursor[0].x,cursor[0].y);
	setbuf(stdout,NULL);
	for(i = 1;i < n;i++)
	{
		sleep(time[i] - time[i - 1]);  //每个点滞留时间为相邻时间节点的差
		printf("\033[%d;%dH ",cursor[i].x,cursor[i].y);
		setbuf(stdout,NULL);
	}
	//消除点
	sleep(4);
	system("clear");
	free(cursor);
	free(time);
	printf("\033[49;30m\033[?25h");
}
```

任务8 修改以上程序，自制一个球体的bmp图片，球体与矩形四边碰到时，会自动反弹，但速度减少10%，当球体停止时，输出球体停止的位置，按任意键程序结束。

任务9 程序调试。

使用了curses.h库中的基本函数来进行简单的图形编程，程序的功能是计算机屏幕呈现一个字符等待终端输入，输入后被另一个字符覆盖，计算每次输入的用时和正确率、平均时间。

```c
#include<sys/time.h>
#include<unistd.h>
#include<stdlib.h>
#include<stdio.h>
#include<curses.h>
int main()
{
struct timeval tv1,tv2;
struct timezone tz;
int f[15];              //每个是否正确
double avrgt,avrgtt,t[15];     //ayrgt 平均时间,avrgtt 反应正确平均时间
double rate,st,stt;     //正确率
int i,s,a;
int c1,c2;
initscr();
printw("It is a typing game, and it will test you reaction time.\nPress any key to start.\n");
getch();
clear();
srand((int)time(0));
while(1){
mvprintw(2,0,"left\taccuracy\ttime\tright?");
for(i=0,s=0;i<15;i++){
 c1='a'+(int)(26.0*rand()/(RAND_MAX+1.0));
 mvaddch(0,0,c1);
 move(1,i);
 gettimeofday(&tv1,&tz);
 c2=getch();
 gettimeofday(&tv2,&tz);
 if(c1==c2){
  f[i]=1;
  s++;
 }
 else
  f[i]=0;
t[i]=tv2.tv_usec-tv1.tv_usec+(tv2.tv_sec-tv1.tv_sec)*1000000;
mvprintw(3+i,0,"%d\t%3.2lf%%  \t%.0lf\t",14-i,100.0*s/(i+1),t[i]);
```

```
    if(f[i])
     printw("right");
    else
     printw("wrong");
   }
   mvaddch(0,0,' ');
   for(i=0,stt=0,st=0;i<15;i++){
    if(f[i]){
     stt=stt+(double)(t[i]);
    }
    st=st+(double)(t[i]);
   }
   avrgt=st/15;
   if(s)
    avrgtt=stt/s;
   else
    avrgtt=0;
   rate=s/15.0;
   mvprintw(3+i,0,"accuracy:%.2lf%%\naverage time:%.1lfms\naverage time to correct:%.1lfms\n",rate*100,avrgt/1000,avrgtt/1000);
   printw("1)quit 2)continue :");
   scanw("%d",&a);
   if(a==1)
    break;
   clear();
  }
  endwin();
  return 0;
}
```

五、实验结果记录

六、实验结果分析

七、实验心得

Linux 程序设计实验报告 14

——虚拟字符驱动程序设计

一、实验目的

1. 掌握在 Linux 环境下设备的分类及设备文件的查看；
2. 理解主设备号与次设备号的概念；
3. 掌握驱动程序编译方法；
4. 掌握设备驱动程序设计流程及驱动程序相关的数据结构；
5. 理解字符设备驱动程序的设计原理；
6. 掌握 Linux 驱动程序的加载、卸载、查看方法。

二、实验内容

1. 查看设备号；
2. 编写程序代码；
3. 模块方法编译；
4. 模块加载与卸载；
5. 虚拟字符驱动程序设计；
6. 驱动程序的模块编译；
7. 设备节点的申请；
8. 驱动程序测试。

三、实验设备

PC 机、Linux 操作系统。

四、实验步骤与操作指导

任务 1 学习字符设备驱动程序框架、驱动程序的编译、加载与卸载。
1. 在 Linux 环境下，设备驱动程序如何划分？
2. 在 Linux 操作系统终端应用命令：ls -l   /dev | more 查看设备驱动程序。

3.在 RedHat Linux 9 的 /home 文件夹下，新建文件夹 driver，用 vi 编辑下面的驱动程序，假定程序名为 hello.c。

```
#include <linux/module.h>
MODULE_LICENSE("Dual BSD/GPL");
#include <linux/init.h>
static int __init hello_init(void)
{
        printk("Hello,world!\n");
        return 0;
}

static void __exit hello_exit(void)
{
        printk("Goodbye,cruel world!\n");
}

module_init(hello_init);
module_exit(hello_exit);
```

4.应用模块加载的方式编译 hello.c

`[root@localhost driver]#gcc -O2 -DMODULE -D__KERNEL__ -c hello.c`

5.加载驱动程序 hello.o

  `insmod hello.o`

注意：

假如出现以下信息：

hello.o: kernel-module version mismatch
hello.o was compiled for kernel version 2.4.20
while this kernel is version 2.4.20-8.

解决方法：

`[root@localhost root]# vim /usr/include/linux/version.h`

将 `#define UTS_RELEASE "2.4.20"`

修改为 `#define UTS_RELEASE "2.4.20-8"`

再次编译，再次加载

6.使用命令 lsmod 查看

Linux 系统中，使用命令 lsmod 查看驱动模块加载的情况与它们的关系。

7.使用命令 cat /proc/modules

8.卸载

  `rmmod hello`

9.查看运行结果

当用 insmod 加载或用 rmmod 卸载驱动时，无法看到 printk 语句的输出，但是，可以从/var/log/messages 中查看到。

```
[root@localhost hello]# cat /var/log/messages |grep world
Feb 26 09:11:23 localhost kernel: Hello,world!
Feb 26 09:34:38 localhost kernel: Goodbye,cruel world!
```

10.修改驱动程序

当用 insmod 加载或用 rmmod 卸载驱动时，若想看到 printk 语句的输出内容，则把 hello.c 文件修改如下：

把"printk("Hello,world!\n");"语句修改为"printk("<0>Hello,world!\n");"

把"printk("Goodbye,cruel world!\n");"修改为"printk("<0>Goodbye,cruel world!\n");"

11.编译

```
[root@localhost hello]# gcc -O2 -D__KERNEL__ -I /usr/src/linux-2.4.20-8/include/ -DMODULE -c hello.c
```

12.加载驱动程序

```
[root@localhost hello]# insmod hello.o
```

显示如下：

```
[root@localhost hello]#
Message from syslogd@localhost at Fri Feb 27 10:08:26 2009 ...
localhost kernel: Hello,world!
You have new mail in /var/spool/mail/root
```

13.卸载驱动

```
[root@localhost hello]#rmmod hello
```

显示如下：

```
[root@localhost hello]#
Message from syslogd@localhost at Fri Feb 27 10:10:58 2009 ...
localhost kernel: Goodbye,cruel world!
```

14.创建设备文件号

```
mknod /dev/hello c 200 0
```

15.应用下述命令查看设备类型、设备属主、主设备号与次设备号

```
ls /dev/hello -l
```

16.编写应用程序进行测试，假定测试程序名 test.c

```c
#include<stdio.h>
#include <sys/types.h>
#include <sys/stat.h>
#include <fcntl.h>
int main()
{
  int testdev;
  char buf[100];
  testdev=open("/dev/hello",O_RDWR);
```

```
    if(testdev==-1)
    {
      printf("Cann't open file\n");
      exit(0);
    }
    printf("device open successe!\n");
}
```

17. 编译 test.c
```
gcc  -o  test  test.c
```
18. 执行 test
```
    ./test
```
19. 如果重写文件 driver.c,重新执行（3）、（4）、（5）、（6）
```
#include <linux/module.h>
#include<linux/init.h>
#include<linux/fs.h>
#define nn  200
struct file_operations gk=
{
};
static int __init hello_init(void)
{       int k;
       printk("Hello,world!\n");
       k=register_chrdev(nn  ,"drive",&gk);
       return 0;
}
Static void __exit hello_exit(void)
{
        Printk ("Goodbye,cruel world!\n");
}


module_init(hello_init);
module_exit(hello_exit);
```

20. 执行查看命令 cat /proc/devices 是否已有 200 driver 一行
21. 执行 test
```
    ./test
```

任务2 嵌入式虚拟字符设备驱动程序设计。

1.字符设备驱动程序结构

如果采用模块方式编写设备驱动程序时，通常至少要实现设备初始化模块、设备打开模块、数据读写与控制模块、中断处理模块（有的驱动程序没有）、设备释放模块和、设备卸载模块等几个部分。下面给出一个典型的设备驱动程序的基本框架，从中不难体会到这几个关键部分是如何组织起来的。

```
/* 打开设备模块 */
static int xxx_open(struct inode *inode, struct file *file)
{
    /*............*/
}
/* 读设备模块 */
static int xxx_read(struct inode *inode, struct file *file)
{
    /*............*/
}
/* 写设备模块 */
static int xxx_write(struct inode *inode, struct file *file)
{
    /*............*/
}
/* 控制设备模块 */
static int xxx_ioctl(struct inode *inode, struct file *file)
{
    /*............*/
}
/* 中断处理模块 */
static void xxx_interrupt(int irq, void *dev_id, struct pt_regs *regs)
{
    /* ... */
}
/* 设备文件操作接口 */
static struct file_operations xxx_fops = {
    read: xxx_read, /* 读设备操作*/
    write: xxx_write, /* 写设备操作*/
    ioctl: xxx_ioctl, /* 控制设备操作*/
    open: xxx_open, /* 打开设备操作*/
    release: xxx_release /* 释放设备操作*/
    /* ... */
```

```c
};
static int __init xxx_init_module (void)
{
  /* ... */
}
static void __exit demo_cleanup_module (void)
{
    pci_unregister_driver(&demo_pci_driver);
}
/* 加载驱动程序模块入口 */
module_init(xxx_init_module);
/* 卸载驱动程序模块入口 */
module_exit(xxx_cleanup_module);
```

2.驱动程序设计环境设置

（1）驱动源程序文件在编译后在/driver 路径下。

（2）编译器为 gcc

因为驱动程序在编译时的工程文件 makefile 中涉及环境设置问题。

3.设计驱动源程序 globalvar.c 文件

```c
#include <linux/module.h>
#include <linux/init.h>
#include <linux/fs.h>
#include <asm/uaccess.h>
MODULE_LICENSE("GPL");
#define MAJOR_NUM 100 //主设备号
static ssize_t globalvar_read(struct file *, char *, size_t, loff_t*);
static ssize_t globalvar_write(struct file *, const char *, size_t, loff_t*);
//初始化字符设备驱动的 file_operations 结构体
struct file_operations globalvar_fops =
{
    read: globalvar_read,         //数据结构中入口函数定义
write: globalvar_write,
};
static int global_var = 0;        //"globalvar"设备的全局变量
static int __init globalvar_init(void)
{
    int ret;
    //注册设备驱动
    ret = register_chrdev(MAJOR_NUM, "globalvar", &globalvar_fops);
    if (ret)
```

```c
        {
                printk("globalvar register failure");
        }
        else
        {
                printk("globalvar register success");
        }
        return ret;
    }
    static void __exit globalvar_exit(void)
    {
        int ret;
        //注销设备驱动
        ret = unregister_chrdev(MAJOR_NUM, "globalvar");
        if (ret)
        {
                printk("globalvar unregister failure");
        }
        else
        {
                printk("globalvar unregister success");
        }
    }
    static ssize_t globalvar_read(struct file *filp,char *buf, size_t len, loff_t *off)
    {
        //将内核空间global_var数据复制到用户空间buf
        if (copy_to_user(buf, &global_var, sizeof(int)))
        {
                return - EFAULT;
        }
        return sizeof(int);
    }
    static ssize_t globalvar_write(struct file *filp, const char *buf, size_t len, loff_t *off)
    {
        //将用户空间的数据buf复制到内核空间global_var中
        if (copy_from_user(&global_var, buf, sizeof(int)))
        {
```

```
        return - EFAULT;
    }
    return sizeof(int);
}
module_init(globalvar_init);
module_exit(globalvar_exit);
```

4.编写驱动程序的 makefile 工程文件

[root@localhost driver]#
gcc -O2 -D__KERNEL__ -DMODULE -c globalvar.c

5.执行 insmod 命令来加载该驱动

/driver # insmod globalvar.o

6.查看/proc/devices 文件，可以发现多了一行 "100 globalvar"

/driver # cat /proc/devices

7.为 globalvar 创建设备节点文件，执行下列命令

/driver # mknod /dev/globalvar c 100 0

创建该设备节点文件后，用户进程通过/dev/globalvar 这个路径就可以访问到这个全局变量虚拟设备了。

8.字符设备驱动应用程序设计

globalvartest.c 文件源代码如下：

```c
#include <sys/types.h>
#include <sys/stat.h>
#include <stdio.h>
#include <fcntl.h>
main()
{
    int    fd, num;
    //打开"/dev/globalvar"
    fd = open("/dev/globalvar", O_RDWR, S_IRUSR | S_IWUSR);
    if (fd != -1 )
    {
        //初次读 globalvar
        read(fd, &num, sizeof(int));
        printf("The globalvar is %d\n", num);
        //写 globalvar
        printf("Please input the num written to globalvar\n");
        scanf("%d", &num);
        write(fd, &num, sizeof(int));
        //再次读 globalvar
        read(fd, &num, sizeof(int));
```

```
            printf("The globalvar is %d\n", num);
            //关闭"/dev/globalvar"
            close(fd);
        }
        else
        {
            printf("Device open failure\n");
        }
}
```

9.编译 globalvartest.c

  [root@localhost driver]#gcc globalvartest.c -o globalvartest

10.应用 globalvartest 测试驱动程序

/driver #   ./globalvartest

五、实验结果记录

六、实验结果分析

七、实验心得

Linux 程序设计实验报告 15

——Linux 串行通信程序设计

一、实验目的

1. 掌握通信端口的打开方法;
2. 理解串口程序设计步骤,掌握串口参数的设置;
3. 掌握串口中字符传送的编程;
4. 掌握串口中文件传送的编程。

二、实验任务与要求

1. 串口的打开方式的程序设计;
2. 串口的设置;
3. 串口通信程序设计;
4. 串口中字符串的传送;
5. 串口中文件的传送。

三、实验工具与准备

PC 机、Linux Redhat Fedora Core6 操作系统。

四、实验步骤与操作指导

任务 1 程序设计。

以 O_RDWR、O_NOCTTY、O_NONBLOCK 方式打开 pc 的 COM1 串行通信端口,并调试。

 提示:

```
int fd;
fd=open("/dev/ttys0",O_RDWR|O_NOCTTY|O_NONBLOCK);
if(fd==-1)
```

任务 2 程序设计。

编写一个串口通信程序,要求使用硬件流控制,8位字符大小,以 9600 的波特率从一台计算机的 COM1 口发送键盘输入的字符,从另一计算机的 COM1 口接收,并在屏幕上打印出接收到的字符。

任务 3 设计一个串口通信程序,通过计算机的 COM1 和 COM2 进行通信,COM1 为发送端,COM2 为接收端,当接收端接收到字符'@'时,另起一段输出;当接收端接收到字符'%'时,结束传输。RS-232 的通信格式为 38400,n,8,1,请分别编写分别编写串口通信中发送端与接收端的 makefile 文件。

源程序代码:

发送端代码:

```c
#include<stdio.h>
#include<stdlib.h>
#include<string.h>
#include<sys/types.h>
#include<sys/stat.h>
#include<fcntl.h>
#include<termios.h>
#define BAUDRATE B38400
#define MODEMDEVICE "/dev/ttyS0"
#define STOP '%'

int main()
{
    int fd,c=0,res;
    struct termios oldtio,newtio;
    char ch,s1[20];
    printf("start...\n");
    fd=open(MODEMDEVICE,O_RDWR | O_NOCTTY);  //打开COM1 端口--发送端
    if(fd<0)
    {   perror(MODEMDEVICE);
        exit(1);
    }
    printf("open...\n");
    tcgetattr(fd, &oldtio);
    bzero(&newtio, sizeof(newtio));
    newtio.c_cflag=BAUDRATE|CS8|CLOCAL|CREAD;
    newtio.c_iflag=IGNPAR;
    newtio.c_oflag=0;
```

```
            newtio.c_lflag=ICANON;
            tcflush(fd,TCIFLUSH);
            tcsetattr(fd,TCSANOW, &newtio);
            printf("writing...\n");
            while(1)
            {
                while((ch=getchar()) != STOP)
            {
                    if (ch != '@'){
                     s1[0]=ch;
                     res=write(fd, s1, 1);
                    }
            else write(fd, "\n\n", 2);
            }
            s1[0]=ch;
            s1[1]='\n';
            res=write(fd, s1, 2);
            break;
                }
        printf("close...\n");
            close(fd);
            tcsetattr(fd,TCSANOW, &oldtio);
            return 0;
}
```

接收端代码:

```
    #include<stdio.h>
    #include<stdlib.h>
    #include<string.h>
    #include<sys/types.h>
    #include<sys/stat.h>
    #include<fcntl.h>
    #include<termios.h>
    #define BAUDRATE B38400
    #define MODEMDEVICE "/dev/ttyS1"
    int main()
    {
            int fd,c=0,res;
            struct termios oldtio, newtio;
            char buf[256];
```

```
        printf("start ...\n");
        fd=open(MODEMDEVICE,O_RDWR | O_NOCTTY);
        if(fd<0)
        {
                perror(MODEMDEVICE);
                exit(1);
        }
        printf("open...\n");
        tcgetattr(fd, &oldtio);
        bzero(&newtio, sizeof(newtio));
        newtio.c_cflag=BAUDRATE|CS8|CLOCAL|CREAD;
        newtio.c_iflag=IGNPAR;
        newtio.c_oflag=0;
        newtio.c_lflag=ICANON;
        tcflush(fd,TCIFLUSH);
        tcsetattr(fd,TCSANOW, &newtio);
        printf("reading...\n");
        while(1)
        {
                res=read(fd, buf,255);
                buf[res] = '\0';
                printf("%s", buf);
                if(buf[0] == '%') break;
        }
        printf("close...\n");
        close(fd);
        tcsetattr(fd,TCSANOW, &oldtio);
        return 0;
}
```

任务4 *程序设计。*

从 COM1 发送字符串到 COM2，在接收端将所有小写字母替换成大写后，再发送到 COM1，并将处理后的字符串打印。提示关键代码：

接收端关键代码：

```
    while(1)
    {
        res=read(fd,buf,255);
        buf[res]=0;/*将读取的信息转换成字符串的格式*/
        if(buf[0]=='@') break;/*这里顺序作了调整*/
```

```
        printf("processing and sending...");
        for(i=0;i<res;i++)
        {
            buf[i]=buf[i]-32;/*小写转为大写*/
            write(fd,buf+i,1);/*发送处理后的字符串*/
        }
}
```

发送端关键代码：
```
while(1)
{
    while((ch=getchar())!='@')
    {
        s1[0]=ch;
        res=write(fd,s1,1);
    }
    res=read(fd,s1,255);/*接受处理后的字符串*/
    s1[res]=0;
    printf("after processed:\n%s\n",s1);
    break;
}
```

任务5 程序调试。

串口传送的应用范围很广，如电报通信、设备基本配置、环境监控、数据采集与交换、工业控制、自动控制等。在自动控制方面，如家电的自动运行、灯光的自动控制、上下班铃声自动控制等给人们的工作生活带来了极大的方便。请编写一个程序，可用菜单方式选择以下操作完成端口间的字符传送、文件的传送、把接收到的内容不仅可显示在屏幕上，还能保存在文件中。思考能否完成图像文件的传送？参考程序中 send1.c 文件和 receive1.c 文件是完成两个终端之间的字符传送，send2.c 文件和 receive2.c 文件是完成两个终端之间文件内容的传送，main 函数来实现用菜单选择方法完成上述操作。

参考代码：

1.main 主程序
```
#!/bin/sh
chmod +x  /root/ send1
chmod +x  /root/ receive1
chmod +x  /root/ send2
chmod +x  /root/ receive2
clear
while true
do
```

```
echo "           Welcome to my program!"
echo "           Wish you to have a good day!"
echo ""
echo "=============================================="
echo "***              ALL FUNCTIONS            ***"
echo "    1 - send the characters                  "
echo "    2 - receive the characters               "
echo "    3 - send the file                        "
echo "    4 - receive & save & show the file       "
echo "    0 - Exit                                 "
echo "=============================================="
echo "please input the number to choose function:"
read num
case $num in
1)./send1;;
2)./receive1;;
3)./send2;;
4)./receive2;;
0)clear
echo "          Thanks for using my program!"
echo "                    Bye-Bye!"
sleep 1
while true
do
exit
done;;
*)echo "please input the right number!"
sleep 1
clear;;
esac
done
```

2.send1.c 源代码如下

```
#include <stdio.h>
#include <stdlib.h>
#include <string.h>
#include <malloc.h>
#include <sys/types.h>
#include <sys/stat.h>
```

```c
#include <fcntl.h>
#include <unistd.h>
#include <termios.h>
#define BAUDRATE B38400
#define MODEMDEVICE "/dev/ttyS0"
#define STOP '@'
int main()
{
int fd,c=0,res;
struct termios oldtio,newtio;
char ch,sl[20];
printf("start…\n");
fd=open(MODEMDEVICE,O_RDWR | O_NOCTTY);
if(fd<0)
{
    perror(MODEMDEVICE);
    exit(1);
}
printf("open…\n");
tcgetattr(fd,&oldtio);
bzero(&newtio,sizeof(newtio));
newtio.c_cflag=BAUDRATE | CS8 | CLOCAL | CREAD;
newtio.c_iflag=IGNPAR;
newtio.c_oflag=0;
newtio.c_lflag=ICANON;
tcflush(fd,TCIFLUSH);
tcsetattr(fd,TCSANOW,&newtio);
printf("writing…\n");
while(1)
{
   while((ch=getchar())!='@')
   {
     sl[0]=ch;
     res=write(fd,sl,1);
   }
     sl[0]=ch;
     sl[1]='\n';
     res=write(fd,sl,2);
     break;
```

```
    }
    printf("close…\n");
    close(fd);
    tcsetattr(fd,TCSANOW,&oldtio);
    return 0;
}
```

3. receive1.c 源代码如下

```c
#include <stdio.h>
#include <stdlib.h>
#include <string.h>
#include <malloc.h>
#include <sys/types.h>
#include <sys/stat.h>
#include <fcntl.h>
#include <unistd.h>
#include <termios.h>
#define BAUDRATE B38400
#define MODEMDEVICE "/dev/ttyS1"
int main()
{
    int fd,c=0,res;
    struct termios oldtio, newtio;
    char buf[256];
    printf("start…\n");
    fd=open(MODEMDEVICE,O_RDWR | O_NOCTTY);
    if(fd<0)
    {
perror(MODEMDEVICE);
exit(1);
    }
    printf("open…/n");
    tcgetattr(fd,&oldtio);
    bzero(&newtio,sizeof(newtio));
    newtio.c_cflag=BAUDRATE | CS8 | CLOCAL | CREAD;
    newtio.c_iflag=IGNPAR;
    newtio.c_oflag=0;
    newtio.c_lflag=ICANON;
    tcflush(fd,TCIFLUSH);
```

```c
    tcsetattr(fd,TCSANOW,&newtio);
    printf("reading...\n");
    while(1)
    {
        res=read(fd,buf,255);
        buf[res]=0;
        printf("res=%d  vuf=%s\n",res,buf);
        if(buf[0]=='@')  break;
    }
    printf("close...\n");
    close(fd);
    tcsetattr(fd,TCSANOW,&oldtio);
    return 0;
}
```

4. send2.c 源代码如下

```c
#include <stdlib.h>
#include <stdio.h>
#include <string.h>
#include <malloc.h>
#include <sys/types.h>
#include <sys/stat.h>
#include <fcntl.h>
#include <unistd.h>
#include <termios.h>
#define max_buffer_size       100       /*定义缓冲区最大宽度*/

int serial_fd;                          /*定义设备文件描述符*/
int open_serial(int k)                  /*k 代表选择的串口*/
{
    if(k==0)
    {
        serial_fd = open("/dev/ttyS0",O_RDWR|O_NOCTTY);
        /*读写方式打开串口*/
        perror("open /dev/ttyS0");
    }
    else
    {
        serial_fd = open("/dev/ttyS1",O_RDWR|O_NOCTTY);
```

```c
        perror("open /dev/ttyS1");
    }
    if(serial_fd == -1)                    /*打开失败*/
        return -1;
    else
        return 0;
}

int main(int argc, char *argv[] )
{
    char sbuf[max_buffer_size];        /*待发送的内容*/
    int retv;
    FILE* filep;
    size_t rsize;
    struct termios opt;
    if(argc!=2)
{
perror("usage: send <filename>\n");
        exit(1);
    }
    if((filep=fopen(argv[1],"r+b"))==NULL)
{
perror("open file error!\n");
        exit(1);
    }
    if(open_serial(0)<0)
{
perror("cannot open ttyS0!");    /*打开串口1*/
        exit(1);
    }
    printf("ready for sending data...\n");   /*准备开始发送数据*/
tcgetattr(serial_fd,&opt);
/*把serial_fd的属性保存到opt中,下面稍作修改*/
    cfmakeraw(&opt);        /*设置某些默认属性*/
    cfsetispeed(&opt,B38400);            /*波特率输出设置为38400bps*/
    cfsetospeed(&opt,B38400);        /*输入设置*/
    tcsetattr(serial_fd,TCSANOW,&opt);      /*马上改变设置*/

    while(!feof(filep))
```

```c
    {
        rsize=fread(sbuf,1,max_buffer_size,filep);
        if(rsize>0)
        {
            retv=write(serial_fd,sbuf,rsize);   /*发数据*/
            if(retv!=rsize)
            {
                perror("write");
            }
            printf("the number of charater sent is %d\n",retv);
        }
    }
    /*当文件发完，发一个终止信号*/
    sbuf[0]='\0';
    retv=write(serial_fd,sbuf,1);
    if(retv!=1)
        perror("sending stop error!");
    if(close(serial_fd)==-1)    /*判断是否成功关闭文件*/
        perror("Close the Device failur!\n");
    if(fclose(filep)<0)
        perror("cann't close the sending file!\n");
    exit(0);
}
```

5. receive2.c 源代码如下

```c
#include <stdio.h>
#include <stdlib.h>
#include <string.h>
#include <malloc.h>
#include <sys/types.h>
#include <sys/stat.h>
#include <fcntl.h>
#include <unistd.h>
#include <termios.h>
#define max_buffer_size  100    /*定义缓冲区最大宽度*/

int fd, flag_close;
char ch;
int open_serial(int k)
```

```c
    {
        if(k==0)        /*串口选择*/
        {   fd = open("/dev/ttyS0",O_RDWR|O_NOCTTY);  /*读写方式打开串口*/
            perror("open /dev/ttyS0");
        }
        else
    {
     fd = open("/dev/ttyS1",O_RDWR|O_NOCTTY);
            perror("open /dev/ttyS1");
        }
        if(fd == -1)   /*打开失败*/
            return -1;
        else
            return 0;
}


int main()
{
    char   hd[max_buffer_size]; /*定义接收缓冲区*/
    int retv,ncount=0;
    struct termios opt;
    FILE* fp;
    /*接收到的数据保存在serialdata文件中*/
    if((fp=fopen("serialdata","wb"))==NULL)
    {   perror("can not open/create file serialdata.");
        exit(1);
    }
    if(open_serial(0)<0)    /*打开串口1*/
    {   perror("open serial port0 fail!");
        exit(1);
    }
    tcgetattr(fd,&opt);
    cfmakeraw(&opt);
    cfsetispeed(&opt,B38400);
    cfsetospeed(&opt,B38400);
    tcsetattr(fd,TCSANOW,&opt);
    printf("ready for receiving data...\n");
    retv=read(fd,hd,max_buffer_size);    /*接收数据*/
/*开始接收数据*/
```

```c
        while(retv>0)          /*判断数据是否接收到*/
        {
            printf("receive data size=%d\n",retv);
            ncount+=retv;
            if(retv>1 && hd[retv-1]!='\0')
                fwrite(hd,retv,1,fp);
/*将接收到的文件内容输入并保存到文件 serialdata 中*/
            else if(retv>1 && hd[retv-1]=='\0')
            {    fwrite(hd,retv-1,1,fp);
/*data end with stop sending signal*/
                break;
            }
            /*单独收到终止信号*/
            else if(retv==1 && hd[retv-1]=='\0')
                break;
            retv=read(fd,hd,max_buffer_size);
        }
printf("The received data size is:%d\n",ncount);
/*print data size*/
printf("\n");
/*将接收到的文件内容输出到屏幕上*/
printf("The content of the file we just received is:\n");
ch=fgetc(fp);
while(ch!=EOF)
{
  putchar(ch);
  ch=fgetc(fp);
}
    flag_close =close(fd);
    if(flag_close== -1)     /*判断是否成功关闭文件*/
        printf("close the Device failur! \n");
    if(fclose(fp)<0)
        perror("closing file serialdata fail!");
    exit(0);
}
```

五、实验结果记录

六、实验结果分析

七、实验心得

Linux 程序设计实验报告 16

——Windows 与 Linux 操作系统间资源共享

一、实验目的

了解 VMware 虚拟机环境下 Linux 与 Windows 之间的资源共享。

二、实验内容

1. 应用VNC与宿主机共享粘贴板；
2. 利用VMware Tools实现Linux与Windows资源共享；
3. 设置虚拟机的共享文件夹，实现Linux与Windows资源共享；
4. 使用Samba服务器实现使用Samba实现共享共享；
5. 在Windows环境下配置GCC编译器；
6. 使用网络将虚拟机中的文件发送到实机。

三、实验设备与准备

PC机、Linux Redhat Fedora Core6操作系统。

四、实验步骤与操作指导

任务1 VNC（虚拟机）与宿主机共享粘贴板。
在 Windows 环境下安装虚拟机，再安装目标主机 Linux，共享粘贴板的设置。
（1）在Linux中执行vncconfig -nowin&；
（2）在Linux选中文字，可直接在Windows中可以粘贴；
（3）在Windows中选中文字，Ctrl+C，在Linux编辑环境中，选择菜单的"粘贴"命令。

任务2 安装 VMware Tools。
实现 VMware 虚拟机中的 Linux 与 Windows 之间的资源共享虚拟机和 Windows 资源的共享给我们带来很大的方便，实现这种资源共享有很多种办法，最主要的办法是使用 VMware Tools 的共享机制。

（1）打开 Vmware 菜单栏 Vm 下的 Setting 中的 Hardware 标签中找到 CD/DVD 那个选项，当然直接在左边那栏开始的时候右键找到或者在右边这个预览框中直接选中也是可以的。

（2）在 CD/DVD 选中后选择右边的 Use ISO Image file 选项，然后找到下面 linux.iso，选择后按[确定]。

（3）加载成功，点击虚拟光盘，把里面 VMwareTools-6.5.0-118166.i368.rpm、VMwareTools-6.0.2-59824.tar.gz 文件包拷到/tmp 目录。

（4）使用命令 rpm -ivh VMwareTools-6.5.0-118166.i368.rpm （或 # tar -zxvf VmwareTools-6.0.2-59824.tar.gz）进行安装（或者可以用图形界面，直接鼠标双击安装）。

（5）进入 cd /usr/bin 目录，找到 vmware-config-tools.pl 文件，运行 ./vmware-config-tools.pl 进行安装。

这样就可以共享了，可以把 Windows 桌面的内容直接拖到 Linux 桌面。

任务3 设置虚拟机的共享文件夹。

（1）在虚拟机关闭的状态下，单击 VM->[setting]->[Options]->[Share Folders]切换到 Options 选项卡,选择 Shared Folders；

（2）然后在右边点击 Add 按键，选择主机目录；

（3）在 Linux 中进到"/mnt/hgfs"就可以看到共享的目录了。

任务4 使用 Samba 实现共享。

Samba 是一个建立在 SMB（Server Message Block）协议上的一个工具套件。它使执行 UNIX 系统的计算机能与执行 Windows 系统的计算机分享驱动器与打印机。非常适用于 Windows 和 Linux 系统并存的网络。与 Samba 服务有关的几个命令如下：

查看 Samba 服务运行状态：# service smb status；

启动 Samba 服务：# service smb start；

停止 Samba 服务：# service smb stop。

（1）首先启动 Samba 服务；

（2）然后在 Windows 系统中设置文件夹共享，并对其设置别名，如：test；

（3）执行命令

smbmount //192.168.0.162/study /mnt/mystudy -o username=admin

其中，192.168.0.162 为 Host 机 IP

（4）在 Windows 中，可以通过打开"运行"，输入 cmd 命令按回车；

（5）查看 ip 地址，在 DOS 环境输入命令 ipconfig(这里的地址是与 Linux 中用 ifconfig 命令得到的属于同一局域网的地址；study 为共享文件夹别名）；

（6）创建挂载点，可由#mkdir /mnt/test 创建；

（7）再打开/mnt/test 目录就能看到共享的文件了。

任务 5 Windows 环境下 GCC 编译器的配置。

Windows 版的 GCC 可以从 MinGW 包中获得，由于 DevC++即使用 MinGW，所以从 DevC++中可以提取到我们所需要的 GCC。用 Windows 版的 GCC 编译程序的方法如下：

（1）安装 DevC++（主要是为了其中的 MinGW）至 C:\Dev-Cpp 目录；

（2）设置环境变量

由于 gcc.exe 是存放在 C:\Dev-Cpp\bin 目录下的，要想在命令行中直接使用 gcc，必须将 C:\Dev-Cpp\bin 加入到系统环境变量 PATH 当中。方法如下：

（A）右击"My Computer"（我的电脑），选择"Properties"（属性），选择"Advanced"（高级）选项卡，点击下方的"Environment Variables"（环境变量），如图：

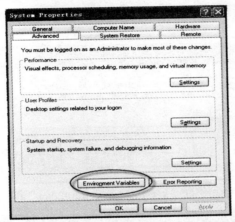

（B）在弹出的对话框中将 C:\Dev-Cpp\bin 加入到 PATH 变量中；

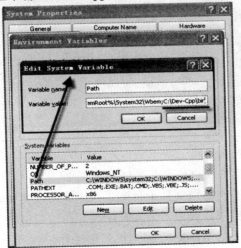

点击 OK，保存设置。

进入 C:\Dev-Cpp\libexec\gcc\mingw32\3.4.5 目录，将 cc1plus.exe 复制一份，新文件名为 cc1.exe。

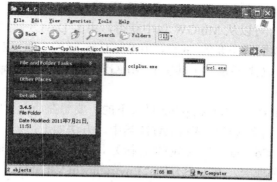

这时候就可以在命令行中使用 gcc 了，按 Win+R 打开"Run"（运行）对话框，输入 cmd 后回车。

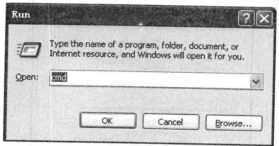

编写好 main.c 后，使用命令行"gcc main.c -o main.exe"即可完成编译。

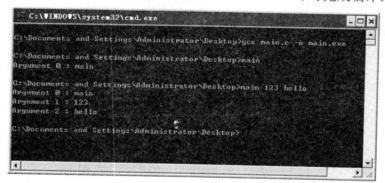

任务 6 使用网络将虚拟机中的文件发送到实机中。

由于 Linux 虚拟机向实机传送文件十分麻烦，因此本次实验将编写两个程序，用于将虚拟机中的文件发送到实机中。其中一个为发送端，在虚拟机的 Linux 中运行，另一个为接收端，在实机（或者另一台 Windows 虚拟机）中运行。

发送端程序在 Linux 环境下以 gcc -o sendfile sendfile.c 编译，发送端程序取名为 sendfile.c，程序代码为：

```
#include <stdio.h>
#include <netinet/in.h>
#include <sys/socket.h>
#include <string.h>
```

```c
#include <netdb.h>
#include <sys/types.h>

#define MAX_LEN 1024 /* 包长 */
#define SIGNATURE 0xAABBCCDD /* 标签,用于验证数据包是否合法 */

int main(int argc, char* argv[])
{
    typedef struct {
            long sig;
            int nSize;
            char szFileName[256];
    } SEND_FILE_HEADER; /* 文件头格式 */
    SEND_FILE_HEADER header;
    int sockfd, recvbytes;
    static char buf[MAX_LEN]; /* 4KB */
    struct hostent* host;
    struct sockaddr_in serv_addr;
    FILE* fp;
    int nSize;

    if(argc != 4)
    {
            printf("Format: sendfile xxx.xxx.xxx.xxx #### filename");
            return -1;
    }

    fp = fopen(argv[3], "rb");
    if(fp == NULL)
    {
            printf("Error: OpenFile ERROR!\n");
    }

    if((host = gethostbyname(argv[1])) == NULL)
    {
            printf("Error: gethostbyname error!\n");
            return -1;
    }
```

```c
    if((sockfd = socket(AF_INET, SOCK_STREAM, 0)) == -1)
    {
        printf("Error: create socket error!\n");
        return -1;
    }

    serv_addr.sin_family = AF_INET;
    serv_addr.sin_port = htons(atoi(argv[2]));
    serv_addr.sin_addr = *((struct in_addr*) host->h_addr);
    memset(&(serv_addr.sin_zero), 0, 8);
    if(connect(sockfd, (struct sockaddr*) & serv_addr, sizeof(struct sockaddr)) == -1)
    {
        printf("Error: Connect error!\n");
        close(sockfd);
        return -1;
    }

    fseek(fp, 0, SEEK_END);
    header.sig = SIGNATURE;
    header.nSize = ftell(fp);
    printf("FileSize: %d\n", header.nSize);
    strcpy(header.szFileName, argv[3]);
    fseek(fp, 0, SEEK_SET);
    memset(buf, 0, MAX_LEN);
    memcpy(buf, &header, sizeof(header));
    send(sockfd, buf, MAX_LEN, 0);

    while(!feof(fp))  /* 循环发送文件 */
    {
        nSize = fread(buf, 1, MAX_LEN, fp);
        if(send(sockfd, buf, nSize, 0) == -1)
        {
            printf("Error: Send Error!\n");
            break;
        }
    }

    close(sockfd);
```

```
    fclose(fp);
    return 0;
}
```

接收端程序在 Windows 环境下用 gcc -o recvfile.exe recvfile.c -lwsock32 编译，程序取名为 recvfile.c，程序代码为：

```c
#include <stdio.h>
#include <winsock2.h>
#define MAX_LEN 1024
#define SIGNATURE 0xAABBCCDD

typedef struct {
    long sig;
    int nSize;
    char szFileName[256];
} SEND_FILE_HEADER;

int main(int argc, char* argv[])
{
SEND_FILE_HEADER header;
static char buf[MAX_LEN];
WSADATA wsa;
char szHost[256];
HOSTENT* pHost;
struct in_addr* pAddr;
FILE* fp;
SOCKET s, sfd;
int nSize = sizeof(struct sockaddr_in);
int i;
struct sockaddr_in sin;
struct sockaddr_in remote_addr;
if(argc != 2)
{
    printf("Format: recvfile portnumber\n");
    return -1;
}
sin.sin_family = AF_INET;
sin.sin_port = htons((unsigned short)atoi(argv[1]));
sin.sin_addr.s_addr = INADDR_ANY;
```

```c
WSAStartup(MAKEWORD(2, 2), &wsa); /* 在 Windows 下，需要先初始化 */
gethostname(szHost, 256);
pHost = gethostbyname(szHost);
if(pHost != NULL)
{
    pAddr = (struct in_addr*)*(pHost->h_addr_list);
    printf("Local IP: %s\n", inet_ntoa(*pAddr));
    s = socket(AF_INET, SOCK_STREAM, IPPROTO_TCP);
    if(s == INVALID_SOCKET)
    {
    printf("Error: Invalid socket\n");
    WSACleanup();
    return -1;
    }
    if(bind(s, (struct sockaddr*)&sin, sizeof(sin)) == SOCKET_ERROR)
    {
    printf("Error: Bind error\n");
    closesocket(s);
    WSACleanup();
    return -1;
    }
    if(listen(s, 5) == SOCKET_ERROR)
    {
    printf("Error: Listen error\n");
    closesocket(s);
    WSACleanup();
    return -1;
    }
    while(1)
    {
    printf("Waiting ...\n");
    if((sfd=accept(s,(struct sockaddr*)&remote_addr, &nSize))==SOCKET_ERROR)
    {
    printf("Error: Accept error\n");
    closesocket(s);
    WSACleanup();
    return -1;
    }
    printf("Accept a link from %s\n", inet_ntoa(remote_addr.sin_addr));
```

```c
recv(sfd, buf, MAX_LEN, 0);
memcpy(&header, buf, sizeof(header));
i = 0;
if(header.sig != SIGNATURE)   /* 验证合法性 */
{
  printf("Signature ERROR! Skip this file\n");
}
else
{
  printf("Receive file: %s SIZE: %d\n", header.szFileName, header.nSize);
  fp = fopen(header.szFileName, "wb");
  if(header.nSize != 0)
  {
    /* 用循环接收文件 */
    while(i < header.nSize)
    {
     if(recv(sfd, buf, MAX_LEN, 0) == SOCKET_ERROR) break;
      if(i + MAX_LEN >= header.nSize)
      {
        fwrite(buf, 1, header.nSize % MAX_LEN, fp);
        i += header.nSize % MAX_LEN;
      }
      else
      {
       fwrite(buf, 1, MAX_LEN, fp);
       i += MAX_LEN;
      }
      printf("\rFinished: %0.2f%%", (double)100.0f * i / header.nSize);
     }
   }
   printf("\rFinished: 100.0%% i = %d\n", i);
   fclose(fp);
  }
 }
}
WSACleanup();  /* 在Windows下，需要清除WSA */
return 0;
}
```

程序运行结果：

以将Linux实验所有源码从Linux发送到Windows中为例。

（1）将本次Linux实验所有源码打包成Experiments.zip：

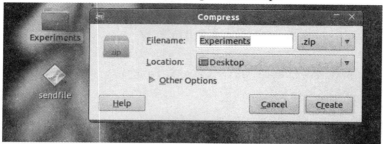

（2）在另一台Windows虚拟机（或实机）中编译并运行recvfile.exe，端口为1234，得到本机IP：192.168.128.128（局域网IP）：

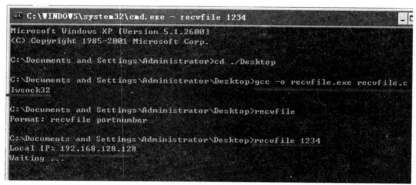

（3）在Linux中，编译并运行sendfile，按命令行"./sendfile 192.168.128.128 1234 Experiments.zip"发送文件。

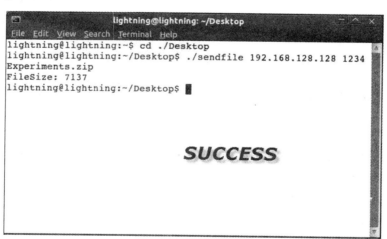

（4）这时在Windows系统中即可看到接收到的文件：

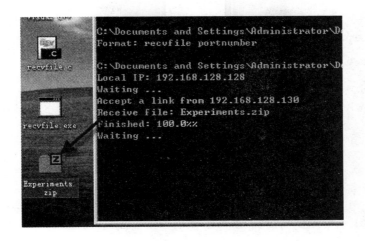

五、实验结果记录

六、实验结果分析

七、实验心得